*Extensional Tectonics
of the Southwestern United States:
A Perspective on Processes and Kinematics*

Edited by

Larry Mayer
Tectonic Geomorphology Laboratory
Department of Geology
Miami University
Oxford, Ohio 45056

© 1986 The Geological Society of America, Inc.
All rights reserved.

Copyright is not claimed on any material prepared
by government employees within the scope of their
employment.

All materials subject to this copyright and included
in this volume may be photocopied for the noncommercial
purpose of scientific or educational advancement.

Published by The Geological Society of America, Inc.
3300 Penrose Place, P.O. Box 9140, Boulder, Colorado 80301

GSA Books Science Editor Campbell Craddock

Printed in U.S.A.

Library of Congress Cataloging in Publication Data

Extensional tectonics of the southwestern United States.

  (Special paper ; 208)
  Includes bibliographies.
  1. Geology—Southwestern States—Congresses.
2. Geology, Structural—Congresses. I. Mayer, Larry,
1951-   . II. Series: Special paper (Geological
Society of America) ; 208.
QE79.5.E98   1986   551.1'3'0979   86-22853
ISBN 0-8137-2208-X

Contents

Preface v

**Topographic constraints on models of lithospheric stretching
of the Basin and Range province, western United States** 1
    Larry Mayer

**Involvement of deep crust in extension of Basin and
Range province** 15
    David A. Okaya and George A. Thompson

Thermal-mechanical consequences of Basin and Range extension 23
    Kevin P. Furlong and Michael D. Londe

**Tertiary structural development of selected basins:
Basin and Range Province, northeastern Nevada** 31
    I. Effimoff and A. R. Pinezich

**Geometry of seismically active faults and crustal deformation
within the Basin and Range-Colorado Plateau transition
in Utah** 43
    Walter J. Arabasz and Dale R. Julander

**Patterns and modes of early Miocene crustal extension,
central Mojave Desert, California** 75
    Roy K. Dokka

**Processes of regional Tertiary extension in the western Cordillera:
Insights from the metamorphic core complexes** 97
    William A. Rehrig

Preface

The study of continental extension in the southwestern United States involves many geologic disciplines and geoscientists whose investigations are conducted at various scales. This Geological Society of America Special Paper is a collection of papers representing a cross-section of those disciplines. Many of these papers represent a new and exciting perspective on rifting within the Basin and Range and Mojave Desert Provinces. They present a combination of modeling and data that illuminate current geologic problems and provide a framework for future research (Table 1).

This volume grew out of a symposium on continental extensional processes held at the combined Cordilleran—Rocky Mountain meeting of the Geological Society of America held in Salt Lake City during May, 1983. That symposium was organized by Roy Dokka (Louisiana State University) and James K. Otton (U.S. Geological Survey). The papers by Dokka, Mayer, and Okaya and Thompson were originally presented at the symposium. The remaining papers contained in this volume were invited in order to present a more complete and comprehensive view of continental extension in the southwestern United States.

The goal of this volume is to provide a better understanding of continental rifting processes in general and in the Basin and Range and Mojave Desert Provinces in particular. Questions of general importance to rifting include: What are the driving forces of rifting? How do initial lithospheric conditions, such as thickness and geothermal gradient, affect subsequent rifting? What are the linkages between plate forces and regional stresses? What structures develop in response to rifting and in which part of the crust do they occur? What are the kinematics of rifting structures? Do different models of lithospheric thinning generate unique geophysical signatures? What are the relations between rifting and regional topography? Do detachment structures represent active rifting structures, or are they relict features?

In addition, certain geologic problems specific to the Basin and Range are addressed by the authors. Why is the Basin and Range Province topographically high given the fact that it has been extended? Because the lithosphere sinks isostatically when buoyant crust is thinned and replaced by denser mantle, the high regional elevation of the Basin and Range is a fundamental problem. Mayer addresses this question in the context of continental rifting. He applies the simple rifting models that have been used to model the subsidence of continental margins to the Basin and Range. The relations among topography, heat, and lithospheric thickness are explored to explain the present and paleotopography of the region.

Given the high elevation of the Basin and Range Province, Okaya and Thompson suggest that addition of large amounts of crustal material or low-density mantle can support high Basin and Range topography and that intrusion of material into the lower crust accompanied Basin and Range extension. Low Pn-velocity (~7.1 km/s) material occurring below the low-density crust and above normal (8 km/s) velocity mantle in many rifts has generally been interpreted as anomalous mantle material. However, this material may be crust that has been intruded by mafic to intermediate composition material resulting in Pn velocities higher than for normal crust. They provide useful conceptual models of the crustal structures that could be associated with extension and intrusion.

Furlong and Londe model the topographic and geothermal signals that would be associated with two specific styles of lithospheric extension; pure shear and simple shear. Differences in topographic and geophysical signatures between styles of extension would provide a tool for determining the relative importance of each style. Their discussion establishes modeling of heat flux as a methodology for evaluating structural styles in continental extensional settings.

Effimoff and Pinezich present seismic reflection data that permit them to describe the geometry of extensional basins in northeastern Nevada. These basins are characterized by listric normal faults that may sole into Mesozoic décollements. They use seismic reflection data as well as surface and well information to outline the structural development of these basins.

The paper by Arabasz and Julander represents an important contribution to the data available on the state of stress and distribution of earthquakes in the Colorado Plateau—Basin and Range transition zone. Their paper reminds us of the absolute need to constrain models with data. Arabasz and Julander present evidence based on detailed earthquake distribution and fault-plane solution data that demonstrate the lack of seismic slip on low-angle and downward flattening faults and also note that the rupture pathways of ground-rupturing earthquakes remain to be determined. Arabasz and Julander point out the apparent contradiction between models that involve high-angle faults connected to low-angle detachments and the data which suggest that surface rupturing earthquakes are probably nucleated at mid-crustal levels (below detachment surfaces). Further, they document

Table 1. Summary of salient aspects of papers in Geological Society of America Special Paper 208

Author	Data Types	Model	Conclusions
Arabasz and Julander	Regional seismic network with supplemental seismic arrays and fault-plane solutions	Relations among crustal structure, geologic structure pricipal stress orientation and earthquake seismology	Predominance of seismic slip on moderately to steeply dipping fault segments. Surface rupture paths of faults nucleating at ca. 15 km depth remain to be identified.
Dokka	Geologic mapping	Kinematics of fault development from structural, stratigraphic data	Extension accomodated by simple shear in upper brittle field above a detachment
Effimoff and Pinezich	Seismic reflection and well data	Conceptual fault models	Faulting is complex. Listric faults active during extension may be related to Mesozoic thrust faults.
Furlong and Londe	Heat flow, gravity and elevation	Geophysical signatures resulting from pure shear and simple shear	Both simple and pure shear models of extension can explain some but not all of the geophysical characteristics across the Colorado Plateau - Basin and Range transition
Mayer	Regional elevation, geomorphology, erosion and sedimentation patterns	Regional elevation and topography from data on crustal and lithospheric thickness	Inflow of low density mantle to compensate for extension and crustal thickening prior to extension
Okaya and Thompson	Gravity, regional elevation, crustal thickness	Elevation and gravity from conceptual fault models	Deep crustal intrusions are coupled with upper crustal extension
Rehrig	Geologic and structure mapping, petrologic and geochemical data	Consequences of pure versus simple shear models on the structural development of extensional orogen and relations to petrology and ore fluid migration	Predominance of pure shear mechanism for extension with secondary simple shear

abundant strike-slip fault-plane solutions in the Sevier Valley region of Utah.

Dokka documents the geometry, kinematics, and ages of extensional structures in the central Mojave desert. In an exemplary field study, Dokka suggests that the high-angle faults and low-angle detachments are kinematically related and, in the Mojave Desert, faulting is preceeded by an interval of dike intrusions.

Rehrig places the development of the Cordilleran metamorphic core complexes in the context of continental extension. The development of the core complexes, specifically the chloritic breccia commonly found at the top of the mylonitic lower plate, may place important constraints on the simple shear model of extension. Rehrig points out that the development of the core-complexes is consistent with a pure shear mode of extension.

This volume would not be possible without the endless patience of the contributors and the high standards of their research. I would like to express many thanks to the authors and especially to George A. Thompson who served as a catalyst for the volume.

Larry Mayer
Oxford, Ohio

Topographic constraints on models of lithospheric stretching of the Basin and Range province, western United States

Larry Mayer
Department of Geology
Miami University
Oxford, Ohio 45056

ABSTRACT

The broad uplift of the Basin and Range province can be described by a two-layer thinning model in which the crustal and subcrustal portions of the lithosphere are thinned by different amounts. Thinning of the lithosphere to the base of the crust can be accomplished by either asthenospheric diapir penetration or by sublithospheric erosion, both resulting in thermal expansion and uplift. The age of crustal extension in the Basin and Range indicates that conduction alone is not likely to be the mechanism for lithospheric heat input because of timing constraints. Application of the two-layer stretching model of Hellinger and Sclater (1983) suggests that the regional topographic features of the Basin and Range province can be largely explained using a sublithospheric thinning factor of $\gamma = 1$ and crustal stretching factors between $\beta = 1$ and $\beta = 1.33$, depending on location. Larger stretching factors are reasonable but may require the assumption of a thickened lithosphere prior to the onset of thinning. Many of the characteristics of the Basin and Range province, such as the location of its eastern boundary, fault geometry, and differences between its northern and southern sections, were strongly influenced by prior tectonic events and are therefore unique to western North America.

INTRODUCTION

Crustal and subcrustal thinning of the Basin and Range province has resulted in a topographic pattern characterized by normal-fault–bounded mountain blocks separated by alluvial valleys that are superimposed on a broad regional uplift. The regional topographic development of the Basin and Range province, as viewed both from present topography and paleotopography inferred from geomorphological studies of erosion and deposition, provides an important kinematic constraint on proposed theories of extensional dynamics and continental rifting. Continental rifting affects the thickness and temperature of crust and subcrustal lithosphere and these changes in turn affect topography. The purpose of this paper is to describe the possible relations among extension, regional topography, and fundamental tectonic features of the Basin and Range province. Many important tectonic features of the Basin and Range province have been inherited from pre-Basin and Range tectonism. Causal relations between lithospheric thinning and topography will be discussed later. The topography of the Basin and Range province is modeled here as a result of two-layer thinning, and comparisons between actual and predicted topography are made. The paper is organized as follows. First I review estimates of extension in the Basin and Range province and describe the tectonic setting in which extension is believed to have developed. The development of tectonic features that predate Basin and Range evolution is then discussed and related to observed features of the present Basin and Range province. Topographic features of the Basin and Range are subsequently described. Finally, rifting models are introduced, and the topography predicted by a two-layer thinning model is compared to the actual topography.

EXTENSION IN THE BASIN AND RANGE PROVINCE

Estimates of crustal extension derived from local studies of fault kinematics are necessary but not sufficient to describe regional lithospheric thinning because only the upper-crustal layer is observed. Extension, E, has generally been reported in the geological literature as a percentage increase in length, i.e., $E = (\Delta l/l)$, where Δl is the increase in length after extension and l is the initial length. However, in the literature on stretching models, either stretching or thinning is preferred (cf. McKenzie, 1978). A stretching factor represented by β is the ratio of the extended length to the initial length, i.e., $(1 + \Delta l)/l$ or simply $\beta = (1 + E)$. Assuming conservation of the extending mass, a thinning factor is defined as $\gamma = 1 - 1/\beta$. Thus, 100% extension is equivalent to a stretching factor of 2.00 or thinning by 50%.

In this paper a distinction is made between pre-Basin and Range extension and Basin and Range extension, though it is the total extension that is of interest in this study. Pre-Basin and Range extension refers to extension that took place in the Basin and Range province during the period ca. 40–15 Ma, before the development of the modern topography. Basin and Range extension and normal faulting is largely responsible for mountain ranges and intervening valleys within the province and therefore much of the modern topography. The distinction between Basin and Range extension and pre-Basin and Range extension is suggested by differences in styles of faulting, inferred regional stress directions, volcanism, and tectonic setting (Zoback and Thompson, 1978; Menges and others, 1981; Shafiquallah and others, 1978; Zoback and others, 1981). An excellent summary of data from which these differences are inferred is given in Zoback and others (1981). Recall, however, that some kinematic interpretations of faulting in the Basin and Range do not require a transition from one type of faulting to another (Wernicke, 1981).

Estimates of upper-crustal stretching in the Basin and Range province are based on geometric interpretations of surface faulting, palinspastic reconstructions based on seismic reflection profiling, palinspastic reconstructions across major strike-slip faults, or interpretation of gravity data in conjunction with mapped faults. Stewart (1971), for example, estimated stretching by assuming the simple geometry of planar faults. Basin and Range crustal stretching (β) has been estimated to be between 1.05 and 1.20 (Hamilton and Myers, 1966; Stewart, 1971; Thompson and Burke, 1974; Zoback and others, 1981). Pre-Basin and Range stretching factors have been estimated to be in excess of 2.00 (Hamilton and Myers, 1966) based on regional palinspastic reconstructions, and similar values have been suggested by Davis (1983) to account for large extensions inferred from the geometry of the tectonite carapace associated with metamorphic core complexes in southern Arizona. Wernicke and others (1982) suggested a minimum $\beta = 1.65$ for stretching in the southern Great Basin portion of the Basin and Range province on the basis of offsets across strike-slip faults.

Models that relate faulting and extension provide the needed link between data accumulated from geologic mapping and stretching estimates. No review of possible fault geometries is given here; the reader is referred to the published literature (for references, see Stewart, 1971; Wernicke, 1981). Okaya and Thompson (this volume), and Furlong and Londe (this volume) provide a geophysical perspective on some faulting models.

The relation between crustal stretching and total lithospheric thinning is an important one. If the entire lithosphere were stretched by a factor of $\beta = 2$, as inferred from data from the upper-crustal layer as discussed above, then one would expect a rapid isostatic collapse of the lithosphere, typical of and analogous to the initial subsidence along a passive margin. Also, if there is uniform stretching of the lithosphere, then we should expect to find some simple connection between stretching estimates and lithospheric thickness. This is not the case, for we find that in some areas in the Basin and Range province, the base of the lithosphere lies essentially at the base of the crust (Thompson and Zoback, 1979), implying extreme thinning of the subcrustal lithosphere, although the Basin and Range province is anomalously high. If the interpretation is correct that in several locations in the Basin and Range the lithosphere lies at the base of the crust, then regardless of assumptions about initial crustal and subcrustal lithospheric thicknesses, it would be impossible to match this interpretation with any type of uniform stretching of the lithosphere. The tectonic setting, within which crustal extension and lithospheric thinning developed, sheds some light on this apparent contradiction. Salient aspects of the tectonic setting are summarized below followed by a discussion on tectonic features in the Basin and Range that were inherited from previous tectonic events.

TECTONIC SETTING OF EXTENSION IN THE BASIN AND RANGE

During much of Paleozoic and Mesozoic time, subduction was occurring along the margin of the western United States. Penultimate convergence and compression-related deformation resulted in the Sevier fold and thrust belt. Final compressional deformation, the initiation of which roughly coincided with a change in plate motions (Coney, 1976), resulted in the Laramide orogeny. The East Pacific Rise collided with the trench about 29 Ma, at a location near Los Angeles, after the subduction of the Farallon oceanic plate (Atwater, 1970). Age estimates for pre-Basin and Range extension (cf. Zoback and others, 1981) range between 40–8 Ma but most commonly fall in the range ca. 30–10 Ma in the northern Basin and Range and ca. 30–17 Ma in the southern Basin and Range province (Chapin and Seager, 1975; Eberly and Stanley, 1978; Shafiquallah and others, 1978).

Two fundamental causal mechanisms have been proposed to explain extension in the Basin and Range. One mechanism calls upon some form of mantle upwelling or diapirism with concomitant thermal incursions into the lithosphere. Mantle-driven extension may have occurred in a back-arc setting (Scholz and others, 1971; Karig and Jensky, 1972; Thompson, 1972; Best and Brimhall, 1974; Thompson and Burke, 1974; Eaton, 1984), over a mantle plume(s) (Suppe and others, 1975), in an intra-arc setting (Coney, 1978), or above a Farallon-slab window (Dickinson and Snyder, 1979).

The second mechanism for extension is oblique shear in response to stresses transmitted from the San Andreas transform system and its consanguine transforms at the boundary between the Pacific and North American plates. This basic hypothesis was proposed by Atwater (1970) in the context of plate kinematics and was subsequently modified or developed by many workers (Heptonstall, 1977; Christiansen and McKee, 1978; Livaccari, 1979). Stresses could be transmitted great distances, presumably due to thermal weakening or softening of the lithosphere. Recent reevaluation of the available data on structural style, paleo- and

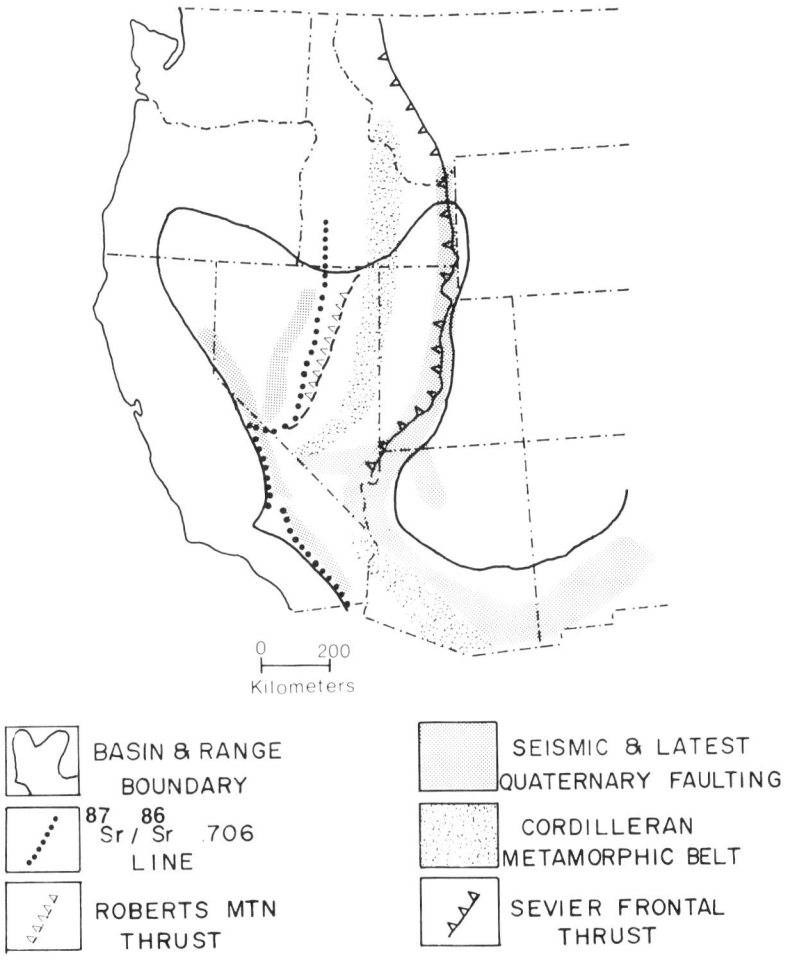

Figure 1. Map showing approximate locations of tectonic features discussed in the text. The location of the belt of active seismicity and latest Quaternary faulting is based upon published earthquake epicenter locations, historical earthquakes and fault ruptures, and age estimates of fault scarps derived from their morphology. The location of the central Nevada seismic belt, for example, is based on historic fault ruptures. The extension of the seismic and fault belt into Arizona is based largely on a few latest Pleistocene(?) fault scarps and not on active seismicity, except for the thumb-shaped protrusion into the Colorado Plateau. The $^{87}Sr/^{86}Sr = 0.706$ line approximates the edge of the Precambrian crust. See text for discussion and references.

modern-stress indicators, and magmatic characteristics (Zoback and others, 1981), led Eaton (1984) to the conclusion that extension in the Basin and Range first developed as an intra-arc, then as a back-arc, and finally due to oblique extension from the San Andreas transform. Of these proposed explanations for extension in the Basin and Range province, the first two provide a way to nonuniformly thin the lithosphere because they involve mechanisms to thin the lithosphere from its base. The latter tectonic mechanism, from an intuitive standpoint, seems to better explain uniform lithospheric extension caused by simple stretching because it involves tensional stresses, presumably with conservation of mass. However, as lithospheric extension and diapiric penetration may be related as cause and effect, it remains possible, if not probable, that multiple processes were operating at the same time rather than in sequence.

INHERITED TECTONIC FEATURES

Fundamental lithospheric characteristics and geologic structures that were developed prior to late Cenozoic extension appear to have strongly influenced the boundaries of the Basin and Range province and its topography, as well as patterns of seismicity (Fig. 1). These geologic features will be discussed in order of their development.

The northern Basin and Range province overlaps the rifted margin of North America that formed in latest Precambrian time, an observation obvious even from cursory examination. Rifting resulted in subsidence, the accumulation of a thick prism of miogeoclinal sediments (Stewart and Poole, 1974; Stewart and Suczek, 1977) west of the present eastern Great Basin boundary, and the development of a thinner cratonal sedimentary sequence to the east, on the present Colorado Plateau. The location of the boundary between thin and thick early Paleozoic sedimentation is located just west of the Colorado Plateau boundary and is now characterized by active seismicity (Smith, 1978). The subsidence history of the Cordilleran miogeocline can be used to determine the amount of rifting-induced lithospheric stretching. A model of simple stretching, proposed by McKenzie (1978) and applied to other rifted margins (Sclater and Christie, 1980; Royden and others, 1980; Le Pichon and Sibuet, 1981; Keen and others, 1981; Sawyer and others, 1982), can be used to estimate the stretching factor needed to account for a subsidence history deduced from stratigraphic data.

In the simple stretching model, the lithosphere is assumed to be stretched by a factor of β resulting in a rapid isostatic collapse of the lithosphere, called the initial subsidence. Subsequent to initial subsidence, cooling and thermal contraction of the lithosphere results in slower subsidence called thermal subsidence. The total subsidence is the sum of initial subsidence, thermal subsidence, and loading by sediments. The stretching factor, β, is estimated by comparing the actual basin subsidence history determined from stratigraphy with the theoretical subsidence history predicted by the McKenzie (1978) model. The comparison is made after removing the effect of sedimentary loading. The procedure of removing that portion of the total subsidence curve due to sedimentary loading is called backstripping (Steckler and Watts, 1978).

The results of backstripping analyses applied to the Cordilleran miogeocline (Fig. 2) and the comparison to the simple stretching model indicate that latest Proterozoic rifting thinned the lithospheric by a factor of about two (Armin and Mayer, 1983; Bond and Kominz, 1984). The thick miogeoclinal sedimentary section in what is now the northern Basin and Range does not have a counterpart in the southern Basin and Range of Arizona, where the lower Paleozoic section is cratonal and significantly thinner (Stone and others, 1983). This thinner sedimentary section implies less lithospheric thinning during latest Proterozoic time in the southern Basin and Range than in the north, and may explain differences in subsequently developed structural characterisics between these areas. In any event, it appears demonstrable that the eastern boundary of the Basin and Range province is located in the area of profound lithospheric contrast caused by latest Proterozoic rifting. The edge of the Precambrian basement is presumably farther west, perhaps best located by the $^{87}Sr/^{86}Sr = 0.706$ isopleth (Kistler and Peterman, 1973; Kistler and others, 1981). Miogeoclinal sedimentation continued uninterrupted until the Antler orogeny (Fig. 2).

The Antler belt extends northward across central Nevada into Idaho and southward into the Mojave Desert (Fig. 1) and generally parallels the Precambrian basement edge. Today this feature is coincident with a line of symmetry separating similar gravity signatures (Eaton and others, 1978), crustal thicknesses (Smith, 1978) and regional topographic lows in the Great Basin. Furthermore, there is a remarkable coincidence between the edge of Precambrian basement and the central Nevada seismic zone as delineated by Wallace (1984).

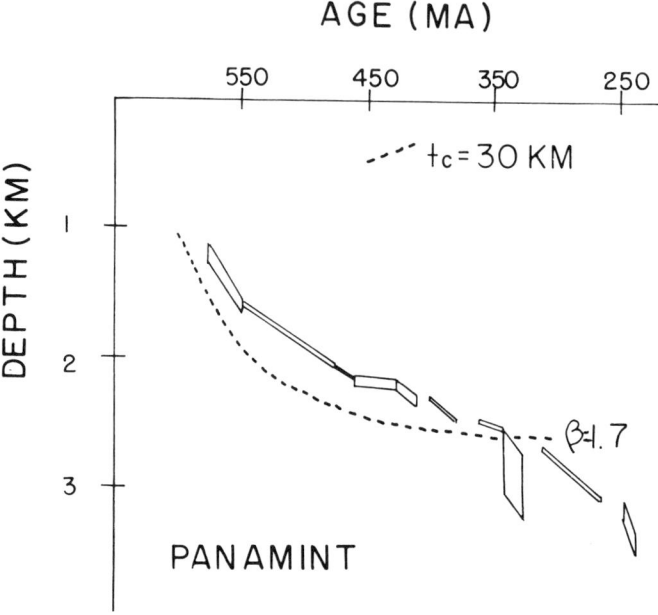

Figure 2. Subsidence history of the Cordilleran miogeocline based on the stratigraphic section preserved in the Panamint Range, California (after Armin and Mayer, 1983). The upper and lower lines represent uncertainties in paleobathymetry. The dashed curve is the thermal subsidence predicted from the uniform stretching model of McKenzie (1978) using a 30-km thick crust.

Mechanisms proposed for the Antler orogeny include continent-arc collision, continent-continent collision, and back-arc thrusting (cf. Nilsen and Stewart, 1980). Speed and Sleep (1982) proposed that an arc system collided with the western margin of North America, resulting in the continentward displacement of the Roberts Mountain allochthon over the Paleozoic shelf and the concomitant formation of a thrust-loading induced foreland basin. Speed and Sleep (1982) suggested that an accretionary prism perhaps 15 km thick is needed to explain the thickness of sediments in the cogenetic foreland basin. The zone of Antler disturbance (±100 km) was subsequently reoccupied by the Sonoma orogeny of Permian-Triassic age (Silberling and Roberts, 1962; though perhaps younger, Ketner, 1984) and by thrusting, such as the Golconda thrust (Silberling and Roberts, 1962; Coats and Riva, 1983) which developed in a back-arc setting (Miller and others, 1984) and the Luning-Fencemaker thrust (Oldow, 1983). Accretion of exotic bits and pieces of arcs, the opening and closing of marginal basins, and other complex plate interactions, including both eastward and westward directed subduction, have been proposed to describe phenomena west of the Precambrian basement edge (Davis and others, 1978; Hamilton, 1978). However, what was to become the western boundary of the present Basin and Range province was not established until Andean-type convergence commenced and the Sierra Nevada batholith was emplaced.

Andean-style convergence at the western margin of North America was characterized by an active volcanic arc in the west

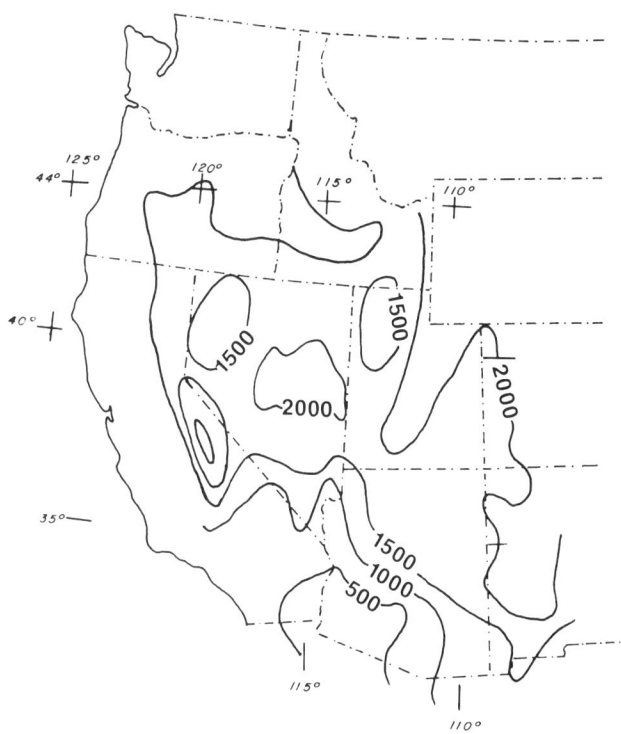

Figure 3. Low-pass filtered regional topographic configuration of the Basin and Range province (modified from Eaton et al., 1978). Topographic contours are in meters above sea level.

and by thin-skinned thrusting of largely unmetamorphosed miogeoclinal and thrust-synchronous foreland basin deposits relatively eastward over little-deformed basement (Armstrong, 1968; Burchfiel and Davis, 1975). West of the thrust belt is the hinterland, characterized by deformation as well as plutonism and metamorphism (Allmendinger and Jordan, 1981). In the eastern Great Basin, the belt of rocks referred to as the Cordilleran metamorphic belt (Miller, 1980) is paired with the Sevier fold and thrust belt.

The eastern Sevier fold and thrust belt is located at the margin of the eastern Great Basin. However, farther south in the southern Basin and Range of Arizona, only the Cordilleran metamorphic belt is found (Fig. 1). The lack of a developed fold and thrust belt is assumed to be due to the nature of the latest Proterozoic rift margin and specifically the lack of a thick miogeoclinal section. Basin analysis of the Sevier thrust-induced foreland basin in central Utah, using backstripping with solutions of elastic flexure, indicates that a 10-km-thick, wedge-shaped thrust load could account for the thickness and geometry of the basin generated by downbending (Lawton and Mayer, 1982). Similar results were obtained by Jordan (1981) for the Idaho-Wyoming thrust belt. The thickness of the thrust load is a conservative estimate of crustal thickening. Another important aspect of Mesozoic compression is the resultant thrust faults that may underlie the northern Basin and Range province. Seismic reflection data indicate the existence of major detachment surfaces such as the Sevier Desert detachment (McDonald, 1976; Allmendinger and others, 1983) in west-central Utah. The Sevier Desert detachment is believed to have accommodated extension as young as Quaternary (Wernicke, 1981) or Pliocene (Allmendinger and others, 1983). Through use of high-resolution reflection profiling, young (Holocene?) fault scarps such as the Clear Lake scarp have been shown to directly connect to the Sevier Desert detachment (Crone and Harding, 1984).

The development of the Basin and Range physiographic province in response to extension began shortly after the close of compressional deformation of the classic Laramide orogeny.

TOPOGRAPHIC CHARACTERISTICS OF THE BASIN AND RANGE PROVINCE

The regional topography of the Basin and Range province provides us with a method to evaluate the usefulness of a particular model of thinning because thinning models explicitly predict changes in the regional topographic surface. The regional topography of the Basin and Range, viewed from low-pass filtered elevation data, shows two distinct sections (Fig. 3). One, the southern Basin and Range, is characterized by a ramplike increase in elevation onto the Colorado Plateau in Arizona and a broad trough centered on the Colorado River. The other section, the northern Basin and Range, is characterized by much higher elevation. The boundary of the northern Basin and Range is most abrupt on the west, adjacent to the Sierra Nevada. On the north, the Snake River Plain is expressed as a tonguelike protrusion into southern Idaho.

The large depression in northwestern Nevada is the Lahontan depression and is characterized by a gravity high (Hildenbrand and others, 1982). The Lahonton depression, also called the Black Rock–Carson Sink zone (Wallace, 1984) is actually rectangular, though this is not visible on the smoothed topography. The second major depression, the Bonneville depression, is located in northwestern–west-central Utah. The boundary between the southern and northern Basin and Range sections is characterized by a steplike increase in elevation and is located at the junction of two strike-slip fault systems: the northwest trending, right-lateral Las Vegas shear zone and the northeast trending, left-lateral Lake Mead shear zone (Bohannon, 1979).

Within the southern Basin and Range of Arizona there are three subprovinces that can be seen on subenvelope maps but not on the low-pass filtered topography. Subenvelope maps are generalized topographic maps produced by contouring only major stream elevations. In southwestern Arizona, the southwestern front of the Growler Mountains defines a boundary separating higher elevations to the east from lower elevations to the west (Fig. 4). This boundary may have subsurface expression farther north along the Gila River (Eberly and Stanley, 1978) and is in the general vicinity of the proposed Jurassic megashear (Silver and Anderson, 1974). Farther east, near Tucson, another topographic transition occurs (Fig. 5) where pronounced differences in texture and relief coincide with a rapid eastward increase in elevation.

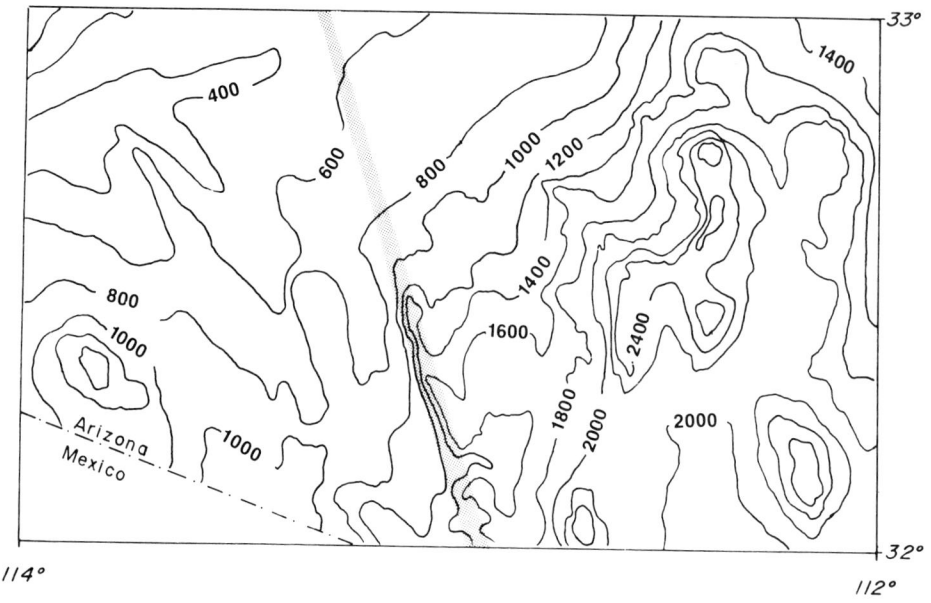

Figure 4. Subenvelope map of the Ajo two-degree topographic sheet (contours in feet above sea level). Most of the individual mountain ranges have little expression. The lineament marked with the stipple pattern separates areas of different topographic texture and elevation. Subenvelope maps are a type of generalized topographic map that filters out local topography by contouring stream elevations.

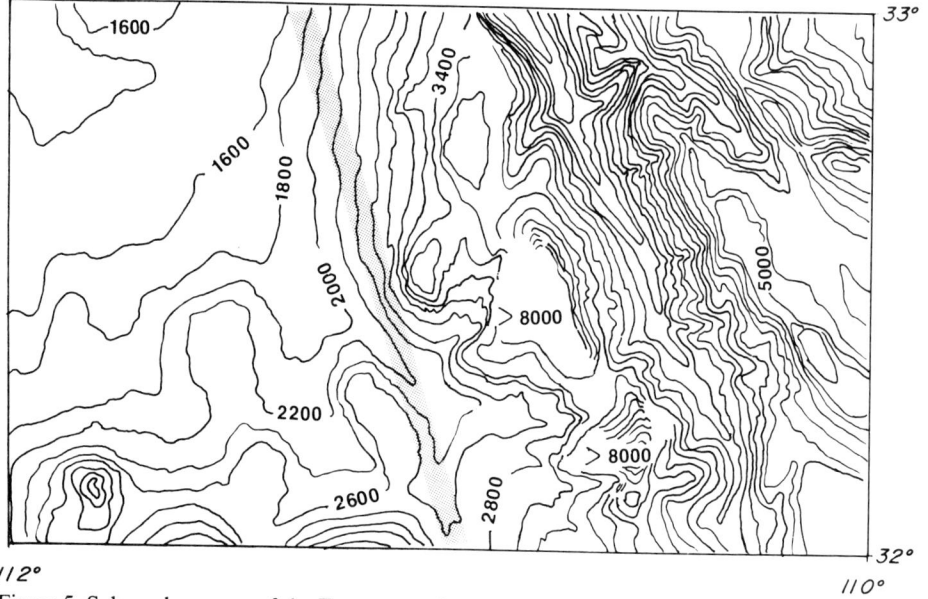

Figure 5. Subenvelope map of the Tucson two-degree topographic sheet (contours in feet above sea level). The lineament, marked with the stipple pattern, separates areas of different average elevation, relief, and topographic texture, and may reflect a crust thicker on the east than on the west. Contours terminate where they were too closely spaced to be shown on the figure.

Topographic transitions between the northern and southern Basin and Range, as well as those from Basin and Range to the Colorado Plateau, can be seen on Figure 6. The Grand Wash Cliffs mark the boundary between the Colorado Plateau and Basin and Range in northwestern Arizona and is topographically abrupt. The boundary between the northern and southern Virgin Mountains marks the transition from northern and southern Basin and Range. The steplike nature of this latter boundary is apparent even at the local scale.

The temporal sequence of Basin and Range topographic development is important in order to evaluate the appropriateness of various proposed rifting mechanisms. The development of the Basin and Range–Colorado Plateau boundary is particularly useful in this regard. Young (1979) has shown that regional north-

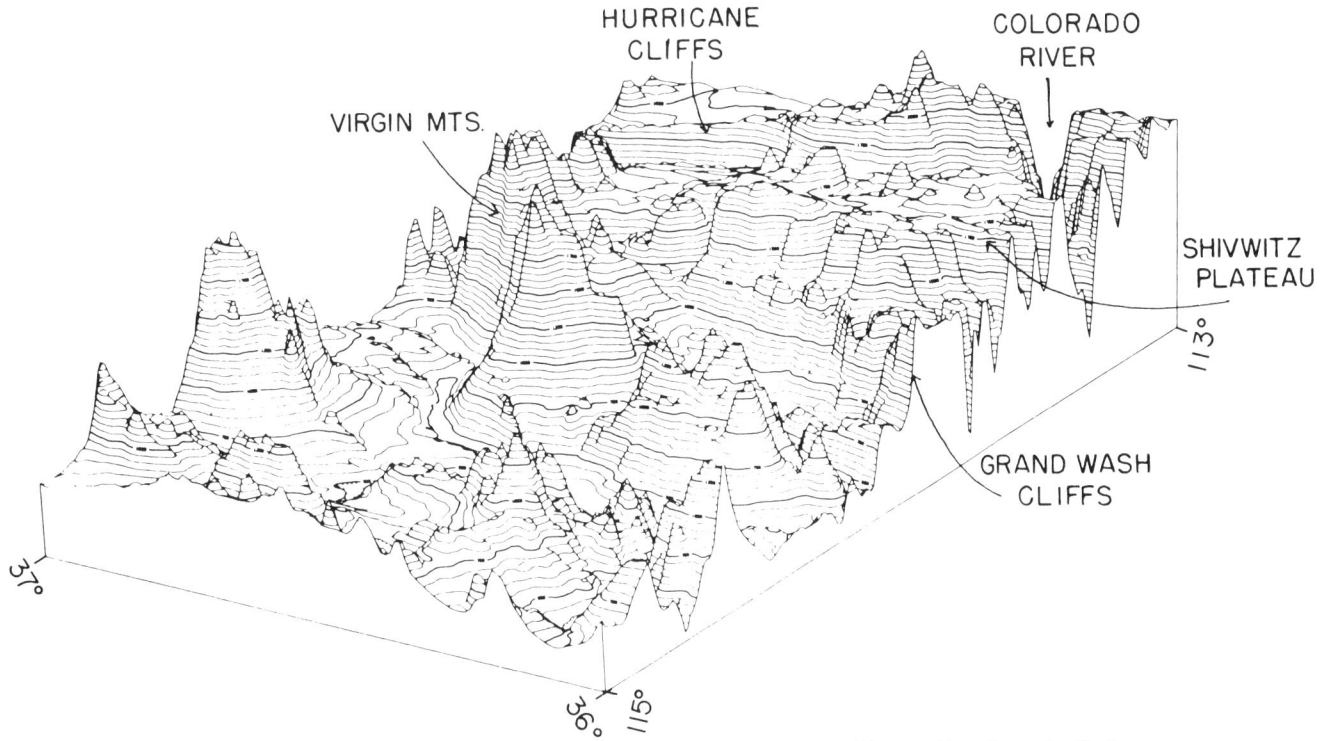

Figure 6. Three-dimensional contour perspective showing the topographic transitions from the Basin and Range province to the Colorado Plateau, and northern to southern Basin and Range province. The Grand Wash Cliffs mark the boundary between the Basin and Range and Colorado Plateau provinces. The decrease in elevation from northern to southern Basin and Range is seen in the left edge of the diagram, which is a topographic profile.

easterly drainage from the Basin and Range onto the Colorado Plateau was initiated during the Laramide and was followed by a period of structural stability, characterized by erosion and subsequent deep weathering. The interval of deep weathering may represent Eocene tectonic quiescence (Young, 1979). The ultimate mechanism of the initiation of the northeasterly drainage pattern may be the topographic highlands caused by crustal thickening in response to Laramide-Sevier compression (as noted above).

The existence of the highlands during mid-Tertiary time is well demonstrated by the production of gravels with a source in the Basin and Range province and their subsequent deposition, by northeasterly-flowing streams, on the Colorado Plateau (McKee, 1951; Cooley and Davidson, 1963; for summary see McKee and others, 1964). The youngest of these gravels in west-central Arizona, the Rim Gravels (Price, 1950), are probably Oligocene in age, ca. 28 m.y. old (Peirce and others, 1979), whereas the oldest gravels deposited after drainage reversal are less than 12 m.y. old (Mayer, 1979; Peirce and others, 1979). However, farther northwest in Arizona, dated deposits indicate that structural differentiation may not have occurred until about 17 Ma (Young and Brennan, 1974). Rowley and others (1978) suggested on the basis of the distributions of ash-flow tuffs in southwestern Utah, that topographic-structural differentiation between Colorado Plateau and Basin and Range may have begun about 26 Ma.

RIFTING MODELS

There are two general types of models of continental rifting; active and passive (Sengör and Burke, 1978). The terms active and passive refer to the role of the asthenosphere in the rifting process. In the passive case, rifting is initiated by regional tensile-stress–induced lithospheric thinning and passive asthenospheric upwelling to replace the thinned lithosphere. The uniform stretching model of McKenzie (1978) is an example of passive rifting, wherein the lithosphere is uniformly stretched; it has been used to explain the subsidence histories of passive margins (see discussion in Inherited Tectonic Features section above). In the active case, rifting is initiated by active asthenospheric upwelling, followed by uplift and extension over the uplift due to lateral spreading (Bott and Kusnir, 1979; Artyushkov, 1973, 1981; Neugebauer and Temme, 1981). Two mechanisms for active-type rifting are (1) thermal thinning of the lithosphere by sublithospheric erosion by which the lower lithosphere is converted to asthenosphere, and (2) asthenospheric diapirs driven by gravitational instability penetrating the lithosphere. Uplift occurs as a result of heating in both active-type mechanisms, and rifting is the result of lateral pressure gradients over the asthenospheric bulge (Artyushkov, 1973). The lateral spreading is considered to be analogous to the ridge-push force (Crough, 1982).

The temporal relations among extension, elevation, and

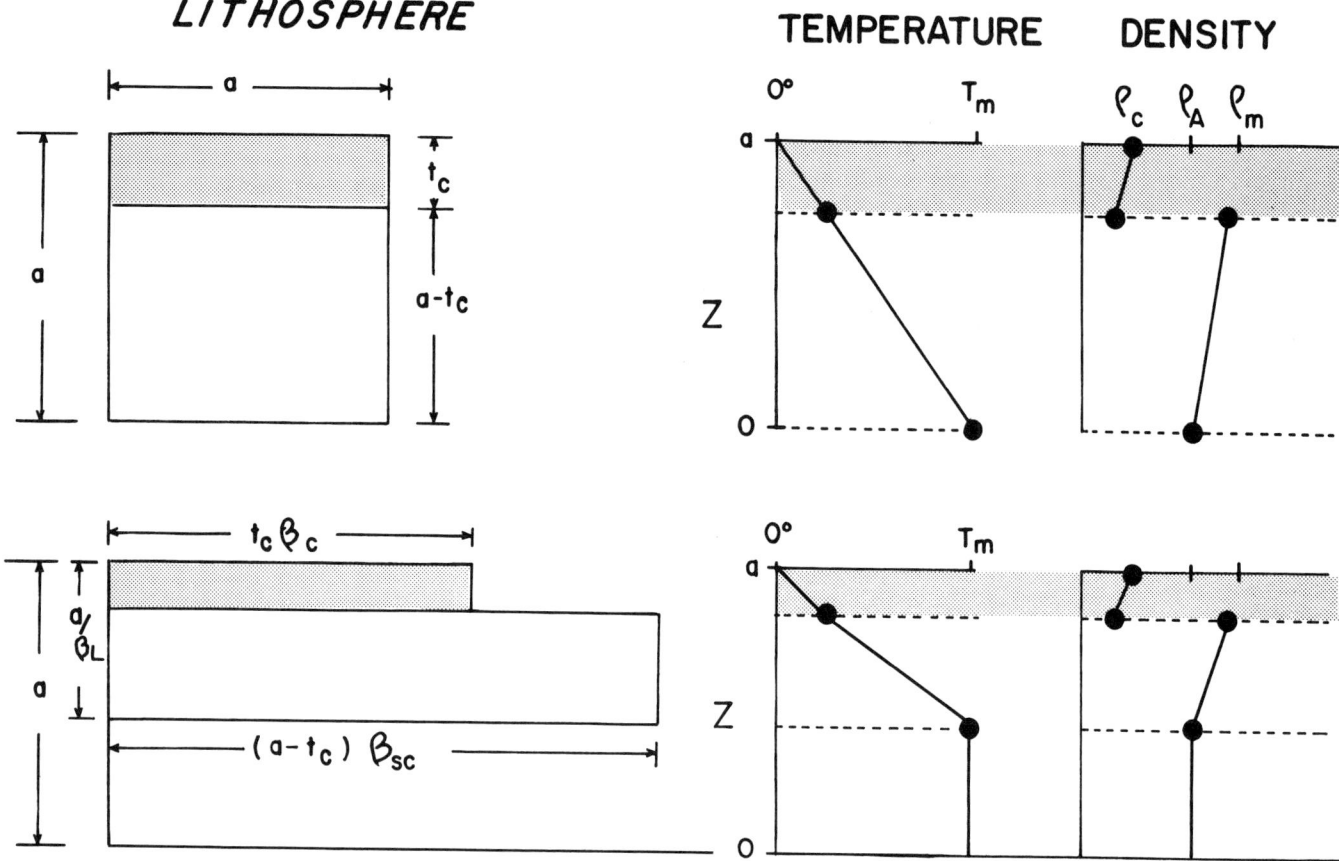

Figure 7. Two-layer stretching model parameters (after Hellinger and Sclater, 1983). Though the model is shown as stretching with mass conservation, the initial isostatic changes are identical without mass conservation. In the latter case, two-layer thinning is a more appropriate description of the model.

lithospheric structure appear to be sensitive indicators in discriminating between active and passive rifting. In the uniform stretching model (McKenzie, 1978), lithospheric extension results in an immediate subsidence followed by continued thermal subsidence. Therefore it is not possible to generate a broad uplift, like that of the Basin and Range province, by uniform extension alone. Also, Sclater and others (1980) and Royden and Keen (1980) found that, for certain passive margin regions, the observed crustal extension did not match the thermal subsidence predicted by uniform stretching. In order to get more thermal subsidence there needed to be more heat put into the lower lithosphere than was allowable, given observed crustal thickness and uniform stretching. The solution was to permit more stretching in the lower lithosphere by using a two-layer stretching model (Sclater and others, 1980; Royden and Keen, 1980; Hellinger and Sclater, 1983). The two-layer stretching model (Fig. 7) predicts uplift where the subcrustal lithosphere is stretched and the crust is not.

In continental rifts where the lithosphere has been thinned to the base of the crust, two-layer stretching poses some room problems (assuming mass conservation) because the extended lower lithosphere has to go somewhere outside of the immediate rift area. If the rift is relatively narrow, a solution to this problem is to distribute the lower lithospheric stretching over some zone wider than crustal stretching. This solution is implicit in the diagrams of two-layer stretching given by Hellinger and Sclater (1983), and mass conservation can thus be maintained. However, where the lithosphere has been thinned to the base of the crust over areas of large extent, mass conservation seems unlikely. Sclater and Royden (*in* Hellinger and Sclater, 1983) noted that mass conservation may not be a necessary condition of the two-layer stretching model. In the latter case, the two-layer stretching model is more appropriately referred to as a two-layer thinning model, and as such may be applied to both active- or passive-type rifting. Turcotte and Emerman (1982) developed a two-layer thinning model for active-type continental rifting.

The regional uplift of the Basin and Range province and adjacent areas implies that active-type continental rifting is involved in extension. The mechanism for the rifting is not clearly understood. Given the plate-tectonic configuration of the region at the time extension is believed to have begun in earnest, it seems reasonable to assume a regional tensile stress regime existed (Atwater, 1970) but that this regional stress was not the cause of extension. This suggests the alternate mechanisms of either asthenospheric diapirism, sublithospheric erosion, or stoping (Mayer, 1983) by a convective system in the upper mantle or hot spot (Suppe and others, 1975). Elevation changes in response to asthenospheric incursions and crustal attenuation can be described by the two-layer thinning model. Elevation changes occur in response to thermal expansion and replacement of buoyant crust with denser mantle. The change in elevation (E) due to thermal expansion alone is (after Sclater, 1972; Royden and Keen, 1980):

$$E = T_m \alpha \mathbf{a} \gamma_L / 2 (1 - \alpha T_m) \quad (1)$$

where $\alpha = 3.3 \times 10^{-5} °C^{-1}$ is the coefficient of thermal expansion, $\mathbf{a}$ is the thickness of the lithosphere, $T_m = 1333°C$ is the temperature at the base of the lithosphere, and γ_L refers to the overall lithospheric thinning. The total change in elevation (S_i), analogous to the initial subsidence in the McKenzie model due to two-layer thinning is given for $S_i > 0$ by (Hellinger and Sclater, 1983):

$$S_i = \{[(\rho_m - \rho_c) t_c (1 - \alpha T_m t_c/2a) - \alpha \rho_i T_m t_c /2] \\ \gamma_c - [\alpha \rho_m T_m (\mathbf{a} - t_c)/2] \gamma_L \} / \rho_m (1 - \alpha T_m) \quad (2)$$

where $\rho_m = 3.33$ g/cm^3 and $\rho_c = 2.8$ g/cm^3 are the densities of mantle and crust at 0°C respectively, t_c is the initial crustal thickness, and γ_c is the crustal thinning factor. Densities are computed as a function of temperature and thermal expansion where the temperature $T = T_m - T_m z/\mathbf{a}$, where z is the distance up to the surface from the asthenosphere–lithosphere boundary. Thus, at base of the crust $T = T_m - T_m (\mathbf{a} - t_c)/\mathbf{a}$ or simply $T_m t_c/\mathbf{a}$. The two-layer thinning model inherently assumes a linear geothermal gradient in the preperturbed state and thus no concentration of radiogenic heat production in the crust. A constant coefficient of thermal expansion for both mantle and crust is also assumed and its value may not be well constrained (Roy and others, 1981, in Morgan, 1982).

Application of the two-layer thinning model to the Basin and Range requires several further assumptions. First, the prethinned lithospheric thickness is assumed to be 120 km, of which 30 km represents the initial crustal thickness. The surface of the 30-km-thick crust is assumed to lie at or near sea-level, given isostatic equilibrium (Watts, 1981, used a 27.7-km-thick crust for consistency with mid-ocean ridge elevation). Second, crustal thicknesses greater than 30 km are assumed to be the remnants of pre-middle Tertiary thickening, perhaps caused during regional compression caused by the Sevier and Laramide orogenies (Kelley, 1955; Corbitt and Wollard, 1970) or some other younger process, and crustal thicknesses less than 30 km are assumed to reflect thinning and or extension younger than 30 Ma. The contribution to the total elevation from crust thicker than 30 km is given by $E_i \approx 0.12 t_c - 3.6$, though this is more important for the areas adjacent to the Basin and Range such as the Sierra Nevada and Colorado Plateau. Similarly, subcrustal lithospheric thicknesses less than 120 km are assumed to be related to thinning. Given these assumptions, the present crustal and lithospheric thickness of the Basin and Range can be used to estimate crustal thinning and lithospheric thinning factors, γ_c and γ_L, respectively.

The compilation of crustal thickness given in Allenby and Schnetzler (1982) was used to estimate crustal thinning factors. Data available on the thickness of the lithosphere are not as detailed as data available on crustal thickness. Available data on depth to asthenosphere were used to compute lithospheric thinning factors, and they suggest a lithosphere 80 km thick under the Colorado Plateau, 110 km under the Sierra Nevada, and at the base of the crust under the Basin and Range (Thompson and Zoback, 1979; Mavko and Thompson, 1983). These data were used in equations (1) and (2) above, and the results are shown as isopleths of thermal uplift and total uplift shown on Figures 8 and 9, respectively. The topography predicted by two-layer thinning is similar to the actual regional topography in form but does not account for all of the observed uplift (cf. Fig. 3 and Fig. 9), perhaps implying a flaw in the assumptions of initial lithospheric configuration or, more specifically, a source of heat still not accounted for. One way to account for this general negative residual is to use an initial lithospheric thickness of 150 km. The results for the thicker lithosphere are shown in Figure 10 and are in better agreement with the observed topography.

MODEL UNCERTAINTIES

The modeled topography shown on Figures 9 and 10 relies heavily on the assumptions of initial lithospheric structure and the value for the coefficient of thermal expansion, as well as the quality of the basic data on crustal and lithospheric thickness. Error limits for the data on crustal thicknesses are not available so I will limit the discussion to model parameters alone. There is no unique way to constrain these values; therefore, it is appropriate to illustrate the effects of variable parameter values in a general way.

Values of the coefficient of thermal expansion commonly cited in the literature on stretching models vary from $3.0-3.3 \times 10^{-5}$ °C^{-1} (Royden and others, 1980) though reasonable values may lie between $3.0 \pm 1 \times 10^{-5}$ °C^{-1} (Molnar and Gray, 1979). The effect on uplift estimates that result from using different values for the coefficient of thermal expansion for given values of $\beta = 1.33$, $\mathbf{a} = 120$ km, and $t_c = 30$ km (Fig. 11) is not trivial but is considered to be limited to less than 200 m.

Errors in estimating the initial crustal thickness (t_c) and lithospheric thickness ($\mathbf{a}$) have large consequences in uplift estimates. There is about 600 m of difference in the uplift predicted assuming a 30-km vs. a 40-km-thick crust (Fig. 12) given $\beta = 1.33$, $\mathbf{a} = 120$, and $\alpha = 3.3 \times 10^{-5}$ °C^{-1}. Varying the value of $\mathbf{a}$ also has a large result on uplift estimates; a 600-m difference in uplift assuming a 100-km vs. a 130-km-thick lithosphere (Fig. 13). However, the differences in uplift estimates obtained by varying crustal and lithospheric thicknesses tend to cancel one another, as long as the initial ratio of crust to total lithosphere is close to the value assumed above (note the slopes of the lines on

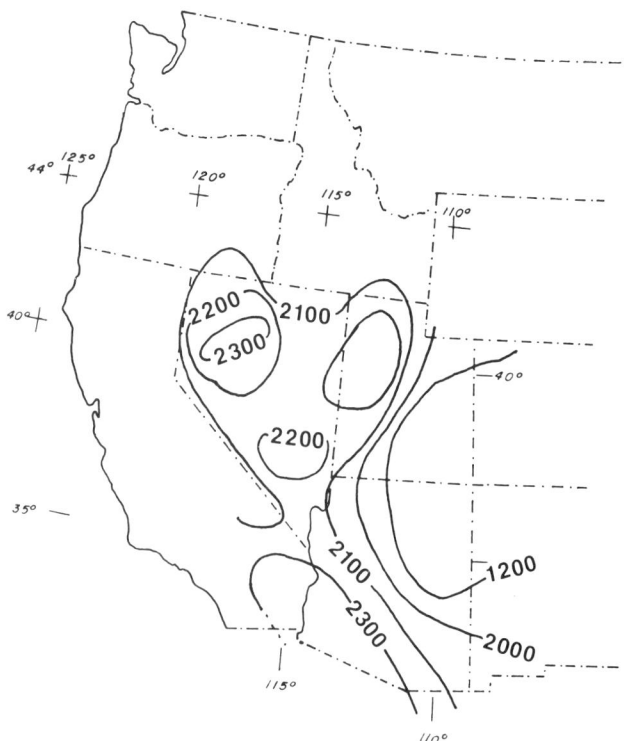

Figure 8. Thermal component of Basin and Range elevation derived from equation (1). If significant thermal incursions preceded crustal thinning, there could have been expressions of this configuration in the paleotopography of the Basin and Range province. For example, northeasterly-flowing streams in Arizona during Eocene and Oligocene time may be a reflection of the northeast sloping surface shown above. Contours in metres. See text for discussion.

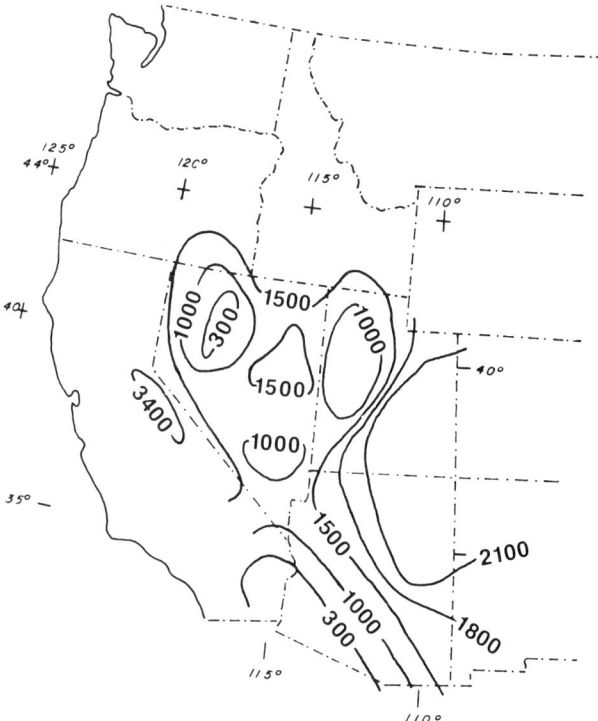

Figure 9. Topography of the Basin and Range province and adjoining areas predicted by the initial topographic changes resulting from two-layer thinning as derived using equation (2). The topography of the Basin and Range province is the sum of thermal uplift shown on Figure 8 and subsidence caused by the replacement of buoyant crust with denser mantle. In the adjoining areas, the topography is the result of thermal uplift and crustal thickening. A 120-km-thick lithosphere with a 30 km crustal portion is assumed as the initial condition. Contours in metres. See text for discussion.

Figs. 12 and 13). The uncertainties in the magnitude of stretching resulting from the above parameter assumptions appear to be most significant in the case where a thickened crust ($>> 30$ km) overlies a nonthickened lithosphere ($\leqslant 120$ km).

DISCUSSION AND CONCLUSIONS

The topographic evolution of the boundary between the eastern Basin and Range–Colorado Plateau implies that prior to extension there was a topographic high in the Basin and Range that permitted drainage onto the plateau. This topographic highland can be interpreted as either early thermal doming or crustal thickening inherited from the Sevier-Laramide orogeny (Wernicke and others, 1982; e.g., see the reconstructed post-Laramide crustal thickness map in Coney and Harms, 1984), or perhaps some combination of both (may be implicit in Haxel and others, 1984). If there was a topographic swell caused by middle Tertiary rifting and thermal expansion, it could not have been caused by heat conducted from the base of the lithosphere because conduction is too slow. Conduction may be a mechanism if heating began sometime earlier, say in the Late Cretaceous. However, Nd and Sr isotopic studies suggest that peraluminous granites younger than 60 Ma were probably generated from crustal melting; i.e., more importantly, melting at mid-crustal levels (Farmer and DePaolo, 1984). The generation of significant melts in the crust may suggest advective heating. The basic mechanism of Basin and Range extension may be asthenospheric diapirism (Smith, 1978).

The magmatism represented by the shallowing Benioff zone during the Laramide (Coney and Reynolds, 1977; Keith, 1978) may provide an important mechanism for the thermal preconditioning of the lithosphere prior to asthenospheric diapirism. The growth and ascent of asthenospheric diapirs into the lithosphere is controlled by the effective viscosity of the lower lithosphere, which is temperature dependent (Bridwell, 1977; Neugebauer, 1982). Heating by an increasingly warm and shallow subducted Farallon plate may be responsible for a reduction in the effective viscosity of the lower lithosphere and the initiation of diapirism sometime during the Paleocene. Bridwell (1977), for example, has proposed asthenospheric diapirism as a mechanism for the Rio Grande rift. The maximum rate of diapiric ascent is on the order of 5 km/m.y. (Neugebauer, 1982) thus requiring at least 20

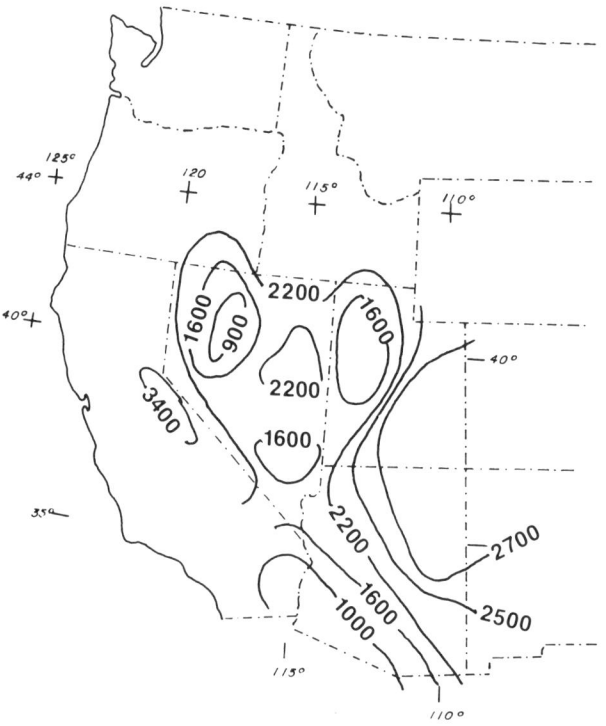

Figure 10. Same as Figure 9 except that the initial lithosphere is assumed to have been 150 km thick. Note by comparing Figures 3, 9, and 10, that the best fit of modeled topography to real topography is obtained by assuming the thicker lithosphere. Contours in metres.

m.y. and perhaps 40 m.y. for the penetration of the subcrustal lithosphere. The latter figure places the initiation of diapirism more closely in time with the cessation of subduction.

The discrepancy between estimates of crustal extension during the period 30–15 Ma and 15 Ma to present is problematic. Using crustal thickness alone yields estimates of $\beta = 1.33$ (assuming 30 km crust initially), in contrast to other estimates of $\beta = 2.00$ (Hamilton and Myers, 1966). A possible solution to this discrepancy is the addition of large amounts of crust by plutonism. Elston (1981) noted that at least one-third of southwestern New Mexico is likely to be underlain by mid-Tertiary plutons. Alternatively, if the initial crustal thickness was closer to 40 km, then there may be a closer agreement between estimates of extension; however, the topography predicted by the two-layer stretching model would deviate even more from the actual topography (accounting for less uplift), perhaps requiring an additional heat source and much greater initial lithospheric thickness. Finally, the low-density rift cushion or rift pillow, seen as anomalously low P_n velocities immediately below the crust under many continental rifts, including the Basin and Range, may have some effect that has not been accounted for in the two-layer thinning model. If the low P_n velocity material at the base of the crust is actually crustal material rather than mantle, it may be easier to integrate Mesozoic crustal thickening with present observations of thinned lithosphere and uplift.

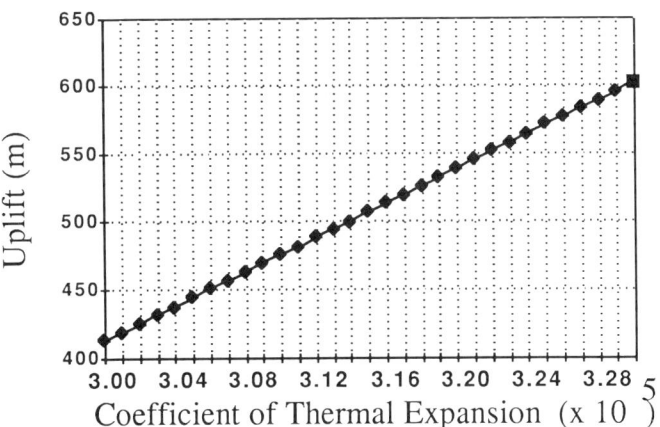

Figure 11. Relation between uplift predicted using equation (2) and assumed values for the coefficient of thermal expansion. All other parameters are fixed as discussed in text. The closed square marks the preferred value.

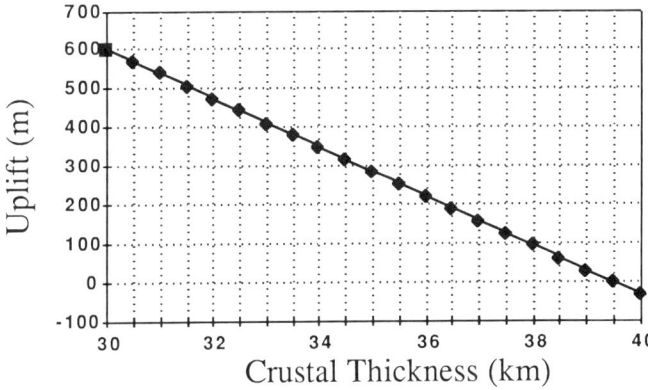

Figure 12. Relation between uplift predicted using equation (2) and assumed initial crustal thicknesses. All other parameters are fixed as discussed in text. The closed square marks the preferred value.

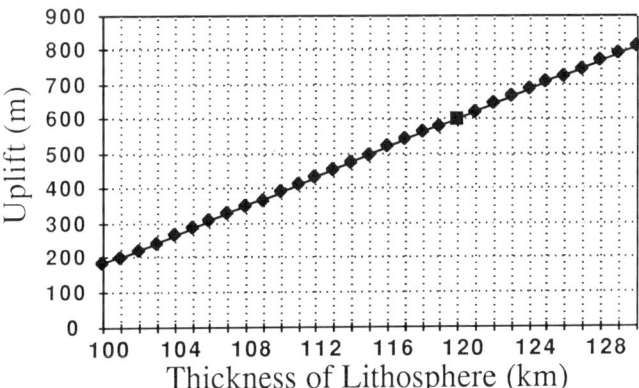

Figure 13. Relation between uplift predicted using equation (2) and assumed initial thickness of the lithosphere. All other parameters are fixed as discussed in text. The closed square marks the preferred value.

In summary, the two-layer thinning model (Sclater and others, 1980; Royden and Keen, 1980; Hellinger and Sclater, 1983) predicts some of the regional topographic features of the Basin and Range province but does not account for all of the observed uplift. Because it is likely that the initial crustal thickness was at least 30 km, a minimum crustal stretching factor of $\beta = 1.33$ is suggested. Given alternate assumptions of a thicker initial lithosphere, it is possible to better match regional topography with predicted uplift. Significant crustal additions cannot be ruled out given the data presented here. Other tectonic features, such as the location of the eastern margin of the Great Basin, are apparently related to late Proterozoic rifting and younger tectonic episodes. The reactivation of older tectonic features in particular seems to be an important characteristic of continental rifts and future models may need to incorporate them. The reactivation of Mesozoic thrust faults as extensional features is an important aspect of regional seismicity. The overlapping in time and space of several tectonic processes makes it difficult, if not impossible, to attribute rifting to a single mechanism. For example, asthenospheric diapirism may have begun ca. 40 Ma in the Basin and Range and later, while the diapirs were ascending, regional tensional stress may have also begun in response to the initiation of the San Andreas transform-fault system. The data summarized above do not negate the notion that several rifting mechanisms were operating simultaneously. More rigorous constraints are needed on rifting models and point to the need for models that more adequately describe the underlying geological processes.

ACKNOWLEDGMENTS

Discussions with W. R. Dickinson and W. Hamilton have helped me define some of the geologic problems of Basin and Range extension that this paper attempts to clarify. I am grateful for the reviews by J. G. Sclater and B. Wernicke.

REFERENCES CITED

Allenby, R. J., Schnetzler, C. C., 1982, United States crustal thickness: Tectonophysics, v. 93, p. 13–31.

Allmendinger, R. W., and Jordan, T. E., 1981, Mesozoic evolution, hinterland of the Sevier orogenic belt: Geology, v. 9, p. 308–313.

Allmendinger, R. W., and 7 others, 1983, Cenozoic and Mesozoic structure of the eastern Basin and Range province, Utah, from COCORP seismic-reflection data: Geology, v. 11, p. 532–536.

Armin, R. A., and Mayer, L., 1983, Subsidence analysis of the Cordilleran miogeocline—Implications for the timing of late Proterozoic rifting and amount of extension: Geology, v. 11, p. 702–705.

Armstrong, R., 1968, Sevier orogenic belt in Nevada and Utah: Geological Society of America Bulletin, v. 79, p. 429–458.

Artyushkov, E. V., 1973, Stresses in the lithosphere caused by crustal thickness inhomogeneities: Journal of Geophysical Research, v. 78, p. 7675–7708.

—— 1981, Mechanisms of continental riftogenesis, in Illies, J. H., ed., Mechanism of graben formation: Tectonophysics, v. 73, p. 9–14.

Atwater, T., 1970, Implications of plate tectonics for the Cenozoic tectonic evolution of western North America: Geological Society of America Bulletin, v. 81, p. 3513–3535.

Best, M. G., and Brimhall, W. H., 1974, Late Cenozoic alkalic basaltic magmas in the western Colorado Plateaus and the Basin and Range transition zone, U.S.A., and their bearing on mantle dynamics: Geological Society of America Bulletin, v. 85, p. 1677–1690.

Bohannon, R. G., 1979, Strike-slip faults of the Lake Mead region of southern Nevada, in Armentrout, J. M., Cole, M. R., and Terbest, H., eds., Cenozoic paleogeography of the western United States: Society of Economic Paleontologists and Mineralogists, Pacific Section, Pacific Coast Paleogeography Symposium no. 3, p. 129–139.

Bond, G. C., and Kominz, M. A., 1984, Construction of tectonic subsidence curves for the early Paleozoic miogeocline, southern Canadian Rocky Mountains: Implications for subsidence mechanisms, age of breakup, and crustal thinning: Geological Society of America Bulletin, v. 95, p. 226–236.

Bott, M.H.P., and Kusnir, N. J., 1979, Stress distribution associated with plateau uplift structures with application to the continental splitting mechanism: Royal Astronomical Society Geophysical Journal, v. 56, p. 451–459.

Bridwell, R. J., 1977, The Rio Grande rift and a diapiric mechanism for continental rifting, in Ramberg, I. B., and Newman, E. R., ed., Tectonics and geophysics of continental rifts: Boston, Reidel, p. 73–80.

Burchfiel, B. C., and Davis, G. A., 1975, Nature and controls of Cordilleran orogenesis, western United States—Extensions of an earlier hypothesis: American Journal of Science, v. 275-A, p. 363–396.

Chapin, C. E., and Seager, W. R., 1975, Evolution of the Rio Grande rift in Socorro and Las Cruces areas, in Seager, W. R., Clemons, R. E., and Callender, J. F., eds.: New Mexico Geological Society, 26th Annual Meeting, Guidebook, p. 297–321.

Christiansen, R. L., and McKee, E. H., 1978, Late Cenozoic volcanic and tectonic evolution of the Great Basin and Columbia intermontane region, in Smith, R. B., and Eaton, G. P., eds., Cenozoic tectonics and regional geophysics of the western Cordillera: Geological Society of America Memoir 152, p. 283–311.

Coats, R. R., and Riva, J. F., 1983, Overlapping thrust belts of late Paleozoic and Mesozoic ages, northern Elko County, Nevada, in Miller, D. M., Todd, V. R., and Howard, K. A., eds., Tectonic and stratigraphic studies in the eastern Great Basin: Geological Society of America Memoir 157, p. 305–327.

Coney, P. J., 1976, Plate tectonics and the Laramide orogeny: New Mexico Geological Society Special Publication no. 6, p. 5–10.

—— 1978, The plate tectonic setting of southeastern Arizona, in Callender, J. F., et al., eds., Land of Cochise: New Mexico Geological Society 29th Annual Meeting, Guidebook, p. 285–290.

Coney, P. J., and Harms, T. A., 1984, Cordilleran metamorphic core complexes: Cenozoic extensional relics of Mesozoic compression: Geology, v. 12, p. 550–554.

Coney, P. J., and Reynolds, S. J., 1977, Cordilleran Benioff zones: Nature, v. 270, p. 403–406.

Cooley, M. E., and Davidson, E. S., 1963, The Mogollon highlands—Their influence on Mesozoic and Cenozoic erosion and sedimentation: Arizona Geological Society Digest, v. 6, p. 7–35.

Corbitt, L. L., and Wollard, C. A., 1970, Thrust faults of the Florida Mountains, New Mexico and their regional tectonic significance: New Mexico Geological Society 21st Annual Meeting, Guidebook, p. 69–74.

Crone, A. J., and Harding, S. T., 1984, Relationship of late Quaternary fault scarps to subadjacent faults, eastern Great Basin, Utah: Geology, v. 12, p. 292–295.

Crough, S. T., 1982, Rifts and swells: Geophysical constraints on causality, in Morgan, P., and Baker, B. H., eds., Processes of continental rifting: Tectonophysics, v. 94, p. 23–37.

Davis, G. H., 1983, Shear-zone model for the origin of metamorphic core complexes: Geology, v. 11, p. 342–347.

Davis, G. A., Monger, J.W.H., and Burchfiel, B. C., 1978, Mesozoic construction of the Cordilleran "collage," central British Columbia to central California, in Howell, D. G., and McDougall, K. A., 1978, eds., Mesozoic paleogeography of the western United States: Society of Economic Paleontologists and Mineralogists, Pacific Section, Pacific Coast Paleogeography Symposium no. 2, p. 1–32.

Dickinson, W. R., and Snyder, W. S., 1979, Geometry of subducted slabs related to San Andreas transform: Journal of Geology, v. 87, p. 609–627.

Eaton, G. P., 1984, The Miocene Great Basin of western North America as an extending back-arc region, in Carlson, R. L., and Kobayashi, K., eds., Geodynamics of back-arc regions: Tectonophysics, v. 102, p. 275–295.

Eaton, G. P., Wahl, R. R., Prostka, H. J., Mabey, D. R., and Kleinkopf, M. D., 1978, Regional gravity and tectonic patterns: Their relation to late Cenozoic epeirogeny and lateral spreading of the western Cordillera, in Smith, R. B., and Eaton, G. P., eds., Cenozoic tectonics and regional geophysics of the western Cordillera: Geological Society of America Memoir 152, p. 51–92.

Eberly, L. D., and Stanley, T. B., 1978, Cenozoic stratigraphy and geologic history of southwestern Arizona: Geological Society of America Bulletin, v. 89, p. 921–940.

Elston, W. E., 1981, Mid-Tertiary extensional orogeny of New Mexico and other parts of the Basin and Range province, in Howard, K. A., Carr, M. D., and Miller, D. M., Tectonic framework of the Mojave and Sonoran deserts, California and Arizona: U.S. Geological Survey Open-File Report 81-503, p. 33–34.

Farmer, G. L., and DePaolo, D. J., 1984, Origin of Mesozoic and Tertiary granite in the western United States and implications for pre-Mesozoic crustal structure. 2. Nd and Sr isotopic studies of unmineralized and Cu- and Mo-mineralized granite in the Precambrian craton: Journal of Geophysical Research, v. 89, p. 10141–10160.

Hamilton, W., 1978, Mesozoic tectonics of the western United States, in Howell, D. G., and McDougall, K. A., 1978, eds., Mesozoic paleogeography of the western United States: Society of Economic Paleontologists and Mineralogists, Pacific Section, Pacific Coast Paleogeography Symposium no. 2, p. 33–70.

Hamilton, W., and Myers, W. B., 1966, Cenozoic tectonics of the western United States: Reviews of Geophysics, v. 5, p. 509–549.

Haxel, G. B., Tosdal, R. M., May, D. J., and Wright, J. E., 1984, Latest Cretaceous and early Tertiary orogenesis in south-central Arizona: Thrust faulting, regional metamorphism, and granitic plutonism: Geological Society of America Bulletin, v. 95, p. 631–653.

Hellinger, S. J., and Sclater, J. G., 1983, Some comments on two-layer extensional models for the evolution of sedimentary basins: Journal of Geophysical Research, v. 88, p. 8251–8269.

Heptonstall, W. B., 1977, Plate linkage mechanism to account for oroclinal deformation in the western Cordillera of North America: Nature, v. 268, p. 763–766.

Hildenbrand, T. G., Simpson, R. W., Godson, R. H., and Kane, M. F., 1982, Digital colored residual and regional Bouguer gravity maps of the conterminous United States: Comments and significant features: U.S. Geological Survey Open-File Report 82-284, 31 p.

Jordan, T. E., 1981, Thrust loads and foreland basin evolution, Cretaceous, western United States: American Association of Petroleum Geologists Bulletin, v. 65, p. 2508–2520.

Karig, D. E., and Jensky, W., 1972, The proto-Gulf of California: Earth and Planetary Science Letters, v. 17, p. 169–174.

Keen, C. E., Beaumont, C., and Boutilier, R., 1981, Preliminary results from a thermo-mechanical model for the evolution of Atlantic-type continental margins: Paris, Oceanologica Acta, Proceedings of the 26th International Geological Congress, 1980, p. 123–128.

Keith, S. B., 1978, Paleosubduction geometries inferred from Cretaceous and Tertiary magmatic patterns in southwestern North America: Geology, v. 6, p. 516–521.

Kelley, V. C., 1955, Regional tectonics of the Colorado Plateau and relationship to the origin and distribution of uranium: University of New Mexico Publication no. 5, 120 p.

Ketner, K., 1984, Recent studies indicate that major structures in northeastern Nevada and the Golconda thrust in north-central Nevada are of Jurassic or Cretaceous age: Geology, v. 12, p. 483–486.

Kistler, R. W., and Peterman, Z. E., 1973, Variations in Sr, Rb, K, Na, and initial $^{87}Sr/^{86}Sr$ in Mesozoic granitic rocks and intruded wall rocks in central California: Geological Society of America Bulletin, v. 84, p. 3489–3512.

Kistler, R. W., Ghent, E. D., and O'Neill, J. R., 1981, Petrogenesis of garnet two-mica granites in the Ruby Mountains, Nevada: Journal of Geophysical Research, v. 86, p. 10591–10606.

Lawton, T. F., and Mayer, L., 1982, Thrust load-induced basin subsidence and sedimentation in the Utah foreland: Temporal constraints on the Upper Cretaceous Sevier orogeny: Geological Society of America Abstracts with Programs, v. 14, p. 542.

LePichon, X., and Sibuet, J-C., 1981, Passive margins: A model of formation: Journal of Geophysical Research, v. 86, p. 3708–3720.

Livaccari, R. F., 1979, Late Cenozoic tectonic evolution of the western United States: Geology, v. 7, p. 72–75.

Mavko, B. B., and Thompson, G. A., 1983, Crustal and upper mantle structure of the northern and central Sierra Nevada: Journal of Geophysical Research, v. 88, p. 5874–5892.

Mayer, L., 1979, The evolution of the Mogollon Rim in central Arizona, in McGetchin, T. R., and Merill, R. B., eds., Plateau uplift mode and mechanism: tectonophysics, v. 61, p. 49–62.

—— 1983, Constraints on non-uniform stretching models of extension in the Basin and Range province: Arizona and adjacent areas: Geological Society of America Abstracts with Programs, v. 15, p. 310.

McDonald, R. E., 1976, Tertiary tectonics and sedimentary rocks along the transition: Basin and Range province to plateau and thrust belt province, Utah: Rocky Mountain Association of Geologists Symposium, p. 281–317.

McKee, E. D., 1951, Sedimentary basins of Arizona and adjoining areas: Geological Society of America Bulletin, v. 62, p. 481–506.

McKee, E. D., Wilson, R. F., Breed, W. J., and Breed, C. S., 1964, Evolution of the Colorado River in Arizona: Museum of Northern Arizona, 67 p.

McKenzie, D., 1978, Some remarks on the development of sedimentary basins: Earth and Planetary Science Letters, v. 40, p. 25–32.

Menges, C. M., Mayer, L., and Lynch, D. J., 1981, Tectonic significance of a post-mid-Miocene stress rotation inferred from fault and volcanic vent orientations in the Basin and Range province of southeastern Arizona and northern Sonora: Geological Society of America Abstracts with Programs, v. 13, p. 96.

Miller, D. M., 1980, Structural geology of the northern Albion Mountains, south-central Idaho, in Crittenden, M. D., Jr., Coney, P. J., and Davis, G. H., eds., Cordilleran metamorphic core complexes: Geological Society of America Memoir 153, p. 399–423.

Miller, E. L., Holdsworth, B. K., Whiteford, W. B., and Rodgers, D., 1984, Stratigraphy and structure of the Schoonover sequence, northeastern Nevada: Implications for Paleozoic plate-margin tectonics: Geological Society of America Bulletin, v. 95, p. 1063–1076.

Molnar, P., and Gray, D., 1979, Subduction of continental lithosphere: Some constraints and uncertainties: Geology, v. 7, p. 58–62.

Morgan, P., 1982, Constraints on rift thermal processes from heat flow and uplift, in Morgan, P., and Baker, B. H., eds., Processes of continental rifting: Tectonophysics, v. 94, p. 277–298.

Neugebauer, H. J., 1982, Mechanical aspects of continental rifting, in Morgan, P., and Baker, B. H., eds., Processes of continental rifting: Tectonophysics, v. 94, p. 91–108.

Neugebauer, H. J., and Temme, P., 1981, Crustal uplift and the propagation of failure zones, in Illies, J. H., ed., Mechanism of graben formation: Tectonophysics, v. 73, p. 33–51.

Nilsen, T. H., and Stewart, J. H., 1980, The Antler orogeny—Mid Paleozoic tectonism in western North America: Geology, v. 8, p. 298–302.

Oldow, J. S., 1983, Tectonic implications of a late Mesozoic fold and thrust belt in northwestern Nevada: Geology, v. 11, p. 542–546.

Peirce, H. W., Damon, P. E., and Shafiquallah, 1979, An Oligocene(?) Colorado

Plateau edge in Arizona, *in* McGetchin, T. R., and Merill, R. B., eds., Plateau uplift mode and mechanism: Tectonophysics, v. 61, p. 1–24.

Price, W. E., Jr., 1950, Cenozoic gravels on the rim of Sycamore Canyon, Arizona: Geological Society of America Bulletin, v. 61, p. 501–507.

Rowley, P. D., Anderson, J. J., Williams, P. C., and Fleck, R. J., 1978, Age of structural differentiation between the Colorado Plateaus and Basin and Range provinces of southwestern Utah: Geology, v. 6, p. 51–55.

Roy, R. F., Beck, A. E., and Touloukian, Y. S., 1981, Thermophysical properties of rocks, *in* Touloukian, Y. S., Judd, W. R., and Roy, R. F., eds., Physical properties of rocks and minerals: New York, McGraw-Hill, 409–502.

Royden, L., and Keen, C. E., 1980, Rifting processes and thermal evolution of the continental margin of eastern Canada determined from subsidence curves: Earth and Planetary Science Letters, v. 51, p. 343–361.

Royden, L., Sclater, J. G., and von Herzen, R. P., 1980, Continental margin and heat flow—Important parameters in formation of petroleum hydrocarbons: American Association of Petroleum Geologists Bulletin, v. 64, p. 173–187.

Sawyer, D. S., Swift, B. A., Sclater, J. G., and Toksöz, M. N., 1982, Extensional model for the subsidence of the northern United States Atlantic continental margin: Geology, v. 10, p. 134–140.

Scholz, C. H., Barazangi, M., and Sbar, M. L., 1971, Late Cenozoic evolution of the Great Basin, western United States as an ensialic interarc basin: Geological Society of America Bulletin, v. 82, p. 2979–2990.

Sclater, J. G., 1972, Heat flow and evolution of the marginal basins of the western Pacific: Journal of Geophysical Research, v. 77, p. 5705–5719.

Sclater, J. G., and Christie, P.A.F., 1980, Continental stretching—An explanation of post-early Cretaceous subsidence of the central graben of the North Sea: Journal of Geophysical Research, v. 85, p. 3711–3739.

Sclater, J. G., Royden, L., Horvath, F., Burchfiel, B. C., Semken, S., and Stegena, L., 1980, The formation of the intra-Carpathian basins as determined from subsidence data: Earth and Planetary Science Letters, v. 51, p. 139–162.

Sengör, A.M.C., and Burke, K., 1978, Relative timing of rifting and volcanism on earth and its tectonic implications: Geophysical Research Letters, v. 5, p. 419–421.

Shafiquallah, M., et al., 1978, Mid-Tertiary magmatism in southeastern Arizona, *in* Callender, J. F., et al., eds., Land of Cochise: Annual Meeting, New Mexico Geological Society, 29th, Guidebook, p. 231–241.

Silberling, N. J., and Roberts, R. J., 1962, Pre-Tertiary stratigraphy and structure of northwestern Nevada: Geological Society of America Special Paper 72, 58 p.

Silver, L. T., and Anderson, T. H., 1974, Possible left-lateral early to middle Mesozoic disruption of the southwestern North American craton margin: Geological Society of America Abstracts with Programs, v. 6, p. 955–956.

Smith, R. B., 1978, Seismicity, crustal structure and intraplate tectonics of the interior of the western Cordillera, *in* Smith, R. B., and Eaton, G. P., eds., Cenozoic tectonics and regional geophysics of the western Cordillera: Geological Society of America Memoir 152, p. 111–144.

Speed, R. C., and Sleep, N. H., 1982, Antler orogeny and foreland basin—A model: Geological Society of America Bulletin, v. 93, p. 815–828.

Steckler, M. S., and Watts, A. B., 1978, Subsidence of Atlantic-type continental margin off New York: Earth and Planetary Science Letters, v. 41, p. 1–13.

Stewart, J. H., 1971, Basin and Range structure: A system of horsts and grabens produced by deep-seated extension: Geological Society of America Bulletin, v. 82, p. 1019–1044.

Stewart, J. H., and Poole, F. G., 1974, Lower Paleozoic and uppermost Precambrian Cordilleran miogeocline, Great Basin, western United States, *in* Dickinson, W.R., ed., Tectonics and sedimentation: Society of Economic Paleontologists and Mineralogists Special Publication no. 22, p. 28–57.

Stewart, J. H., and Suczek, C. A., 1977, Cambrian and latest Precambrian paleogeography and tectonics in the western United States, *in* Stewart, J. H., et al., eds., Paleozoic paleogeography of the western United States: Society of Economic Paleontologists and Mineralogists, Pacific Section, Pacific Coast Paleogeography Symposium no. 1, p. 1–17.

Stone, P., Howard, K. A., and Hamilton, W., 1983, Correlation of metamorphosed Paleozoic strata of the southeastern Mojave Desert region, California and Arizona: Geological Society of America Bulletin, v. 94, p. 1135–1147.

Suppe, J., Powell, C., and Berry, R., 1975, Regional topography, seismicity, Quaternary volcanism and the present day tectonics of the western United States: American Journal of Science, v. 275-A, p. 397–436.

Thompson, G. A., 1972, Cenozoic Basin and Range tectonism in relation to deep structure: Proceedings International Geological Congress, 24th, Montreal: Quebec, Harpell's Press, p. 84–90.

Thompson, G. A., and Burke, D. B., 1974, Regional geophysics of the Basin and Range province: Annual Review of Earth and Planetary Science, v. 2, p. 213–238.

Thompson, G. A., and Zoback, M. L., 1979, Regional geophysics of the Colorado Plateau, *in* McGetchin, T. R., and Merill, R. B., eds., Plateau uplift mode and mechanism: Tectonophysics, v. 61, p. 149–181.

Turcotte, D. L., and Emerman, S. H., 1982, Mechanisms of active and passive rifting, *in* Morgan, P., and Baker, B. H., eds., Processes of continental rifting: Tectonophysics, v. 94, p. 39–50.

Wallace, R. E., 1984, Patterns and timing of late Quaternary faulting in the Great Basin province and relation to some regional tectonic features: Journal of Geophysical Research, v. 89, p. 5763–5769.

Watts, A. B., 1981, The U.S. Atlantic margin—Subsidence history, crustal structure and thermal evolution: American Association of Petroleum Geologists Educational Course Note Series no. 19, 75 p.

Wernicke, B., 1981, Low-angle faults in the Basin and Range province: Nappe tectonics in an extending orogen: Nature, v. 291, p. 645–648.

Wernicke, B., Spencer, J. E., Burchfiel, C., and Guth, P. L., 1982, Magnitude of crustal extension in the southern Great Basin: Geology, v. 10, p. 499–502.

Young, R. A., 1979, Laramide deformation, erosion and plutonism along the southwestern margin of the Colorado Plateau, *in* McGetchin, T. R., and Merill, R. B., eds., Plateau uplift mode and mechanism: Tectonophysics, v. 61, p. 25–47.

Young, R. A., and Brennan, W. J., 1974, Peach Springs tuff—Its bearing on structural evolution of the Colorado Plateau and development of Cenozoic drainage in Mohave County, Arizona: Geological Society of America Bulletin, v. 85, p. 83–90.

Zoback, M. L., and Thompson, G. A., 1978, Basin and Range rifting in northern Nevada—Clues from a mid-Miocene rift and its subsequent offsets: Geology, v. 6, p. 111–116.

Zoback, M. L., Anderson, R. E., and Thompson, G. A., 1981, Cainozoic evolution of the state of stress and style of tectonism of the Basin and Range province of the western United States: Royal Society of London, Philosophical Transactions, v. A-300, p. 407–434.

MANUSCRIPT ACCEPTED BY THE SOCIETY MARCH 4, 1986

Involvement of deep crust in extension of Basin and Range province

*David A. Okaya**
George A. Thompson
Department of Geophysics,
Stanford University,
Stanford, California 94305

ABSTRACT

Although geologic and reflection seismic studies are making rapid strides in defining the geometry and indicating the processes involved in upper crustal extension of the Basin and Range province, fundamental processes in the deeper crust and the upper mantle are still poorly constrained. We summarize interrelated gravity, elevation, isostatic, and geologic evidence to suggest that the convective mass circuit of rifting is completed partly within the crust by intrusion of material of crustal density from the mantle. If this interpretation is correct, mechanical thinning of the crust by extension, especially in highly extended terranes of the Basin and Range province, is somewhat reduced by additions of new crust. Conceptual models for links between upper crustal faulting-stretching and deeper crustal igneous dilation are presented. The deeper crust has probably been profoundly changed by metamorphic reaction and partial melting.

INTRODUCTION

Faulting and ductile spreading in rift zones that we can observe and map are shallow phenomena compared to the thickness of the lithosphere. We know this because gravity data indicate unequivocally that the mass removed by geologically observed extension is replaced by an inflow of mass at depths of no more than a few tens of kilometres. For example, extensional thinning of only the upper ten kilometres of crust (as known in exposures and drilling in some rift zones) by 20% would result in a local decrease of the isostatic gravity anomaly by about 200 mgal (the attraction of the missing two-kilometre "plate") if compensating mass did not flow in at greater depth. Local gravity anomalies of this magnitude are not observed in rifts, and those that are observed are generally associated with rock masses of anomalous density, such as low-density sedimentary fill. This type of convective mass circuit is comparatively well known where thin lithospheric plates diverge at ocean ridges, although details such as the depth of backflow in the asthenosphere and the shapes of shallow magma chambers are unresolved.

In continental rifts we observe brittle extension of the upper crust by faulting and, where exposures of sufficient depth permit, ductile extension of the middle crust, as in metamorphic core complexes. Commonly, however, the volume of volcanic and exposed intrusive rocks that accompanied extension, as well as sedimentary rocks deposited during extension, is grossly inadequate to make up the mass that has been transported away laterally by spreading in the shallow crust. In some rifts igneous rocks are sparse or absent; for example, as in the western branch of the East African Rift system (in contrast to the eastern branch or rift) and in parts of the Basin and Range province. However, we argue that large igneous additions to the deeper crust are likely. The geophysical evidence in the Basin and Range province derives mainly from gravity, regional elevation, and crustal thickness, and is consistent with other observations such as heat flow.

An alternative interpretation, often stated or implied, is that the crust simply rose isostatically during or following extension. Such an interpretation requires an inflow of dense mantle material beneath a crust thinned by extension. This process would leave highly extended terranes far below the elevations where we find them; in fact, generally below sea level. The point of considering both the gravity data and the land elevation is that gravity tells us about the inflow of mass, and elevation tells us something about the density of that mass. For example, gravity across the Mid-Atlantic Ridge (Talwani and others, 1965) indicates that isostatic compensation is taking place; i.e., that the mass deficit

---
*Present address: CALCRUST, c/o Earth Science Division, Lawrence Berkeley Laboratories, Berkeley, California 94720.

created by the diverging lithospheric plates is replaced by inflow of an equal amount of asthenospheric mass. Moreover, the elevation of the ridge, about three kilometres above the ocean basin, indicates that the inflowing mass has a lower density than the lithosphere that it replaces. Similarly, highly extended continental areas that have not subsided to or below sea level may have experienced an inflow of material of crustal density or of anomalously low mantle density. If, additionally, the crust has not been thinned by the requisite amount for the geologically measured extension, then the case is strong for inflow of material of crustal density. The most likely source of new crust is basalt and andesite from the mantle.

What processes couple the brittle faulting of the upper crust with deformation and intrusion in the deep crust? In this paper we examine alternative geometrical models that are consistent with the evidence.

GEOLOGIC BACKGROUND—BASIN AND RANGE PROVINCE

The tectonic development of the Basin and Range province has been extensively reviewed (e.g., Zoback and others, 1981; Eaton, 1982) and need only be briefly summarized here. Within the province are regions of extreme extension (highly extended terranes), some of which are termed metamorphic core complexes (e.g., Proffett, 1977; Gans and Miller, 1983). The highly extended terranes vary in their age of development but are generally early to middle Cenozoic. They are associated with arc-type magmatism and in any given locality are older than the block faulting that produced the present basins and ranges. They have been called a product of intra-arc spreading. These terranes are characterized by closely spaced (1 km), strongly tilted (>30°) fault blocks, like a laid-over shelf of books. In some areas of deep exposure (metamorphic core complexes), these blocks can be seen to rest on subhorizontal detachment faults which are underlain by ductilely extended rocks (e.g., Miller and others, 1983).

In contrast, the faulting and tilting responsible for the present Basin and Range topography are generally younger than those that produced the extreme extension in any given area, and the fault blocks are much wider (10 km) and less tilted. Thus the highly extended terranes are usually exposed within the younger uplifted blocks. The younger faulting is characteristically associated with basaltic (or bimodal basalt-rhyolite) volcanism.

Although the style, dimensions, degree of tilting, and volcanic associations of the highly extended terranes are in contrast with the younger Basin and Range structure, the possibility remains that much of the difference may be attributable to the degree of extension and to the generally shallower structural level at which we are able to observe young Basin and Range structures. Specifically, seismic reflection data indicate that high-angle faults bounding modern ranges bottom in subhorizontal (ductile?) detachment faults at depths of less than 20 km (Allmendinger and others, 1983; Effimoff and Pinezich, 1981; McDonald, 1976; Smith and Bruhn, 1984). Angelier and Colletta (1983) have suggested that with increasing amounts of extension, large Basin and Range blocks evolve into small, highly tilted blocks.

LINK WITH THE DEEP CRUST

In the Basin and Range province, evidence and interpretation indicate that the faulted, seismogenic upper crust gives way at depths of about 5–15 km to ductilely extended deeper crust. Evidence includes: (1) the depth of earthquakes (maximum about 20 km) and the effect of temperature on crustal rocks (e.g., Bott, 1980; Sibson, 1982; Eaton, 1982); (2) observations on deeply exposed crustal sections in metamorphic core complexes (e.g., Coney, 1980; Davis and others, 1980; Davis, 1980, 1983; Miller and others, 1983); and (3) seismic reflection sections that show high-angle normal faults bottoming in subhorizontal detachment faults (e.g., McDonald, 1976; Effimoff and Pinezich, 1981; Allmendinger and others, 1983; Cape and others, 1983). Evidence is contradictory on the change in character of normal faults with depth. Data on the larger historical earthquakes in the Basin and Range province (Okaya and Thompson, 1985; Richens and others, 1985; Stein and Barrientos, 1985) clearly indicate that these normal faults dip about 60° and are planar to a depth of about 15 km, which is the hypocentral depth and the maximum aftershock depth. The focal mechanisms, aftershock distributions, geodetic strains, and geologic observations support this interpretation. On the other hand, reflection seismic records show that many normal faults in the Basin and Range province are listric, i.e., they gradually flatten with depth in the style of growth faults, and become subhorizontal at depths of 5–10 km (e.g., Anderson and others, 1983, Cape and others, 1983; Smith and Bruhn, 1984). A possible resolution of this contradiction is that the planar faults are primary and are associated with more deeply penetrating zones of crustal deformation. The listric faults, on the other hand, may be secondary breaks along which gravitationally driven blocks move laterally toward the space created by the primary zones, somewhat like landslides. We were forced to this concept of primary and secondary deformation (illustrated diagramatically in Figure 5) while analyzing geophysical data from the Rio Grande rift (Cape and others, 1983). Moore (1960) advocated a mechanism of this type based upon the spoonlike shape of many Basin and Range faults. Are we to conclude from the foregoing evidence that the whole crust, or perhaps the whole lithosphere, has been extended by about the same amount, the upper crust by brittle faulting and the lower crust by ductile stretching? We think, instead, that igneous dilation of the crust has increased with depth.

The difference between igneous additions to the crust and the normal isostatic process of mantle inflow is in the density of the added material. Material of crustal density (such as basalt) can be extracted from nonresidual mantle peridotite without increasing the density of the peridotite. This happens because of the mineral phase changes involved in the differentiation process; the net volume of basalt plus the residual peridotite increases, and the mean density decreases. We do not mean to imply that the intrusions are all basalt. More silicic magmas (such as andesites) may be derived directly from a subducted plate in the mantle, or basalt rising into the crust may react with and partially melt the crust to create more silicic intrusions (Hildreth, 1981; Taylor, 1980; Hamilton, 1981; Lachenbruch and Sass, 1978). Lachenbruch and Sass showed that the high heat flow in the Basin and Range province is consistent with intrusion models.

Geologic evidence for igneous additions to the crust includes the widespread occurrence of basalts, as well as earlier andesites and ash-flow tuffs, in the Basin and Range province and Rio Grande rift. Basalt dikes surely tapped the mantle and therefore extend through the whole crust. One would expect basalt to be trapped beneath lower density rocks of the upper crust and only rarely to find its way to the surface. Moreover, the rise of silicic magmas is inhibited by their viscosity. Bacon's (1982) studies in the rifted Coso area of California indicate that only a small fraction of the total igneous input to the crust reaches the surface. Plausibly, large volumes of magma came into the lower crust in extended regions and tended to maintain crustal thickness during brittle extension and thinning of the upper crust. This type of model has been advocated for extension in the Basin and Range province by Rehrig and Reynolds (1980), Gans and Miller (1983), Thompson and Burke (1974), and Thompson (1959). The model has the great advantage of preserving crustal thicknesses now observed (25–30 km) even in areas of extreme extension (more than 200%), and of maintaining high, isostatically supported elevations. We clearly recognize, however, that the upper mantle in the Basin and Range province is anomalous and that the land can be elevated regionally by density reductions in the mantle. A more complete understanding of the relative roles of the crust and the mantle will have to await seismic detailing of lateral variations in both the crust and the upper mantle.

Models

Simplified conceptual models illustrate the problem of transition between upper crustal faulting and deeper crustal extension. We discuss a range of models as a framework for examining needed tests and currently available constraints. Figure 1 illustrates alternative tilt-block geometries and Figure 2 illustrates a single graben without tilt. In Figure 1A, the base of the blocks was assumed to merge into ductilely extended deeper crust. In Figure 1B, an early listric-fault model, Evison (1960) postulated that only the upper crust (3 km) of the Basin and Range province was undergoing extension. We now know, from the lack of complementary zones of thrusting or compression, that the deep crust must also be involved in extension. Figures 2A and 2B show geometrically acceptable links with the deeper crust; these models are equally applicable to the tilt blocks of Figure 1. Let us assume that the deep crust is ductilely extended on the average by the same amount as the upper crust and examine the isostatic consequences.

In any region of extending fault blocks such as those in Figures 1 and 2, if the deeper crust is also extending and if we

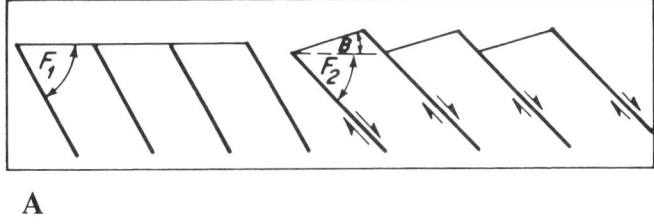

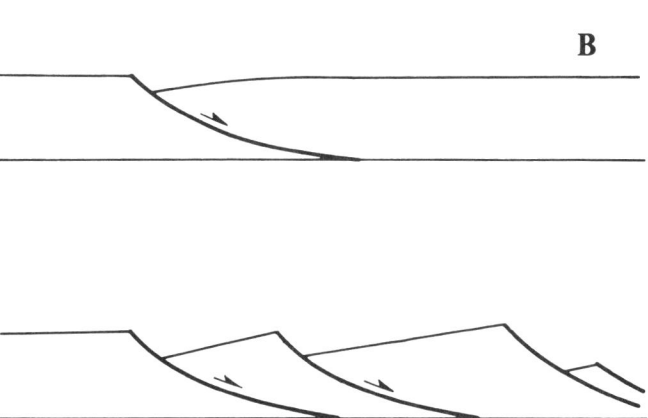

Figure 1. Two diagrams that illustrate a long-standing contrast in interpretation of extensional style. A: Tilted blocks; F_1 and F_2 are dips of faults before and after tilting and B is dip of bedding; the percent extension is $100 (\sin F_1/\sin F_2 - 1)$. The base of the tilted blocks is assumed to be accommodated by more complex fracturing and a gradual transition to ductile deeper crust, at a depth of about 10 km (Thompson, 1960). B: Two stages in an early model of listric Basin and Range faulting (Evison, 1960).

consider a region broad enough (50–100 km) to respond isostatically, the mean elevation will decrease. We here neglect mechanical thinning of mantle lithosphere, if it occurs, as a second-order effect because of the small density contrast between lithosphere and asthenosphere (~ 0.05 g/c^3) compared to that between crust and mantle (~ 0.4 g/c^3). The amount of subsidence depends upon the original thickness, the density contrast between crust and mantle, and the extension. For example (Fig. 3), a crust 42 km thick (like the present Colorado Plateau), with a density contrast of 0.4 g/c^3 with the mantle, if extended and thinned 50% (to 28 km thick), would subside 1.7 km.

Because the Basin and Range province stands at an average elevation of about a kilometre or more above sea level, and the most intensely extended parts of it are not notably lower, low-density mass must have moved into the system instead of high-density mantle as implied by ordinary isostatic adjustment such as that shown in Figure 3. Magma differentiated from the mantle and intruded into (added to) the crust is the likely explanation.

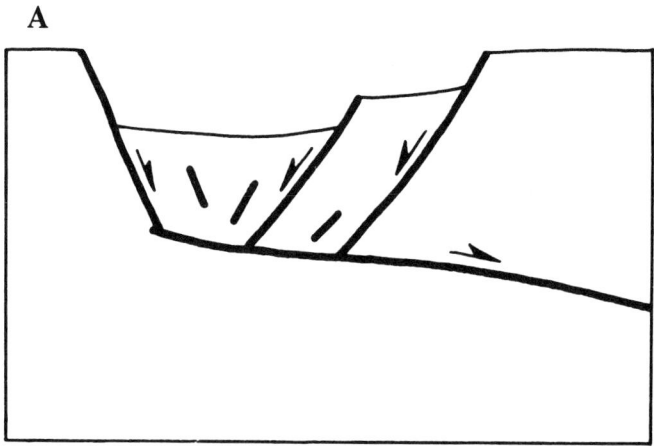

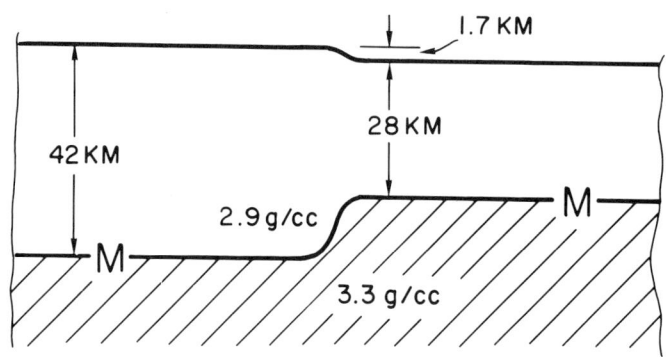

Figure 3. Simplified model of crust extended and thinned 50% (right side); isostatic compensation leaves the altitude 1.7 km below its original height.

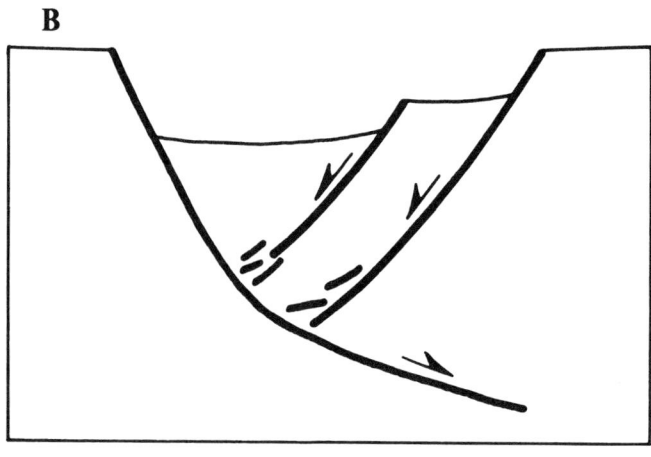

Figure 2. Geometrically plausible models for the transition to deeper crust below a graben whose shallow structure is known (Dixie Valley, Nevada; about 10 km wide). A: Based on the geometry of the known upper crustal structure, the low-angle detachment would commence in the brittle upper crust (<10 km depth) and extend into the ductile lower crust or into the mantle. B: A "master" high-angle fault penetrates the ductile lower crust, possibly shallowing in the process. Antithetic normal faults root in this master fault.

Figures 4A and 4B illustrate conceptually how intrusions might be related directly or indirectly to a graben.

Figure 5 (Cape and others, 1983) is a sketch of the shallow fault structure of the Rio Grande rift derived from seismic reflection data and a postulated deeper structure suggested by crustal earthquakes and by magma bodies detected seismically (Sanford and others, 1977; Brown and others, 1980).

To further quantify these conceptual models, Figure 6 shows a computer-generated model of extension in the wedge-shaped upper plate of a low-angle normal fault as conceived by Wernicke (1981), except that we have chosen to flatten the fault to subhorizontal in the ductile lower crust instead of extending it to the asthenosphere. Figure 6A shows that amount of upper crustal extension by the overlap at the right-hand side of the diagram; we think the lower part of the crust would be extended also and its thickness maintained by intrusions. The computed Bouguer gravity anomaly is also shown in Figure 6A; no isostatic compensation is assumed. After compensation (Fig. 6B), which in nature would occur during development of the structure, the topographic surface is slightly lower than originally, the mantle is upwarped, and the Bouguer and isostatic gravity anomalies have returned to values expected for a region of that elevation that is in isostatic equilibrium.

The model shown in Figure 6 maintains the thickness of the lower crust and in this respect is equivalent to our intrusion model. If the low-angle fault had penetrated (and thus thinned) the entire crust, or if the lower crust had been thinned ductilely, the topographic surface would have been lowered much more, as in Figure 3, and the local positive Bouguer anomaly would have been much greater. For the elevation to be maintained, or even decreased, the mean-crustal density would have to be decreased (for example, by intrusions), or the upper mantle density would have to be decreased. In view of the fact that highly extended terranes (core complexes) are commonly exposed in mountain ranges and generally do not have large positive Bouguer anomalies associated with them, intrusion models are strongly favored over simple fault models such as those of Wernicke (1981).

DISCUSSION AND CONCLUSIONS

In continental rift zones, as in oceanic rifts, material beneath the crust flows laterally toward and into the rifted regions as the upper, brittle plates move apart, thus completing the convective circuit. A major question in continental rifts is how the crust participates in this process. Is it merely attenuated and finally pulled apart, to be replaced by mantle and a thin oceanic crust, or is the crust at first replenished by infusions of material of crustal

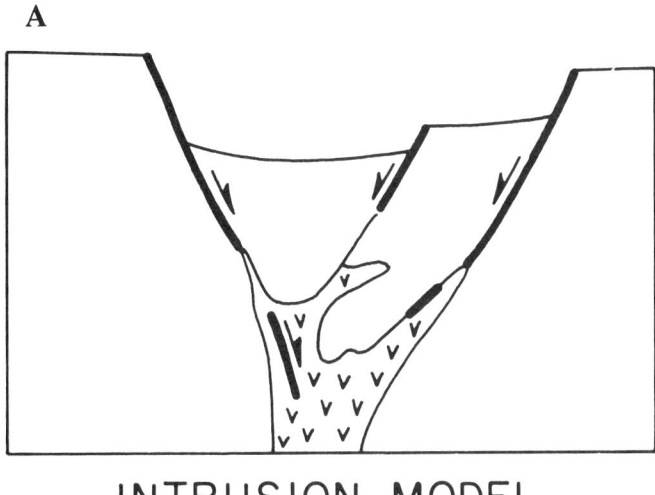

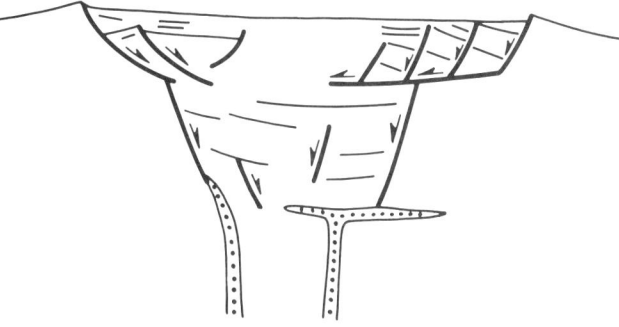

Figure 5. Diagrammatic crustal section of Rio Grande rift after Cape and others (1983). The upper part, to a depth of about 6 km, shows the tilt blocks and detachment faults indicated by seismic reflection data; the lower part, which is hypothetical, shows steeper faults or fractures suggested by earthquakes and by young basaltic volcanism. Magma has been detected by both earthquake (Sanford and others, 1977) and reflection (Brown and others, 1980) seismology. Although not to scale, the diagram represents the whole crust.

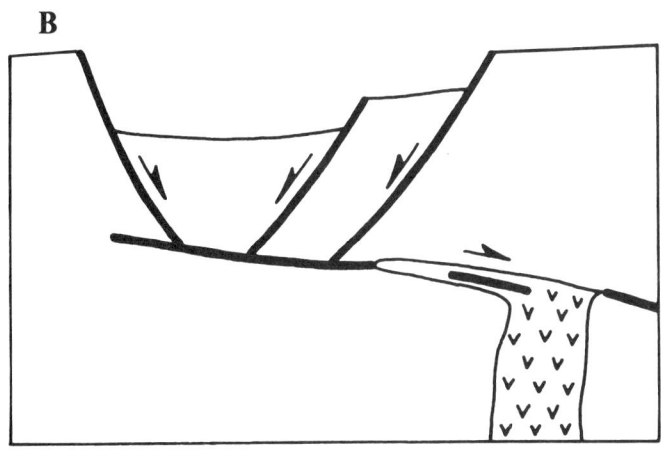

Figure 4. Intrusion models in which the crustal thickness below a graben such as that in Figure 2 is maintained or partially maintained by igneous additions; regional altitude can be isostatically maintained, in contrast to Figure 3. A: Graben-bounding faults root into lower to mid-crustal intrusions. B: Graben-bounding faults root into a mid-crustal low-angle detachment which in turn roots into intrusions.

density derived from the mantle? We have argued from general principles and regional evidence that the latter process (crustal additions) is important in Basin and Range spreading.

Seismic refraction and reflection studies of the crust and upper mantle structure are supplying new tests of spreading models and may soon elucidate the fundamental process. Recently, Mooney and others (1983) presented new data for the Mississippi embayment and made intriguing comparisons with other continental rift zones, including the Salton trough, Rhinegraben, Gregory rift, and Rio Grande rift. What these and other continental rifts seem to have in common is a pillow-like mass of high-velocity (7.2-7.5 km/s) basalt crust. The high-velocity material is inferred to be lower crust modified by injection of material from the mantle during rifting.

We note here that injected material with velocities in the range of normal crust would not likely be revealed seismically. Therefore, a considerable additional part of the crust above the high-velocity basalt pillow may consist of intrusions. The way in which modified lower crust might have been formed is not discussed by Mooney and others (1983). We suggest that injection of basalt into the crust may have been accompanied by reaction melting (Taylor, 1980; Barker and others, 1975). This process would tend to concentrate the most refractory, residual, high-velocity material in the lowermost crust, while material that rose higher in the crust could have velocities within the normal crustal range. It is also worth noting that some of the rifts having high-velocity lower crust are young, active, and hot. Therefore, there is no interpretation involving high-pressure, low-temperature granulites in the lower crust.

Although the Rio Grande rift is the only part of the Basin and Range tectonic province in which the anomalous high-velocity lower crust has been reported, high-resolution refraction studies, especially in the areas of greatest extension, are sparse. Moreover, upper-mantle velocities as low as 7.4 km/s have been reported in the eastern Basin and Range (Keller and others, 1975); the question of what is crust and what is mantle is bound to arise in this circumstance.

In summary, we suggest that evidence strongly favors substantial intrusions into the crust during the process of Basin and Range extension, and we have offered a few conceptual models of the link between intrusions and upper crustal faulting. It also seems likely that intrusive basalt reacts with and partially melts

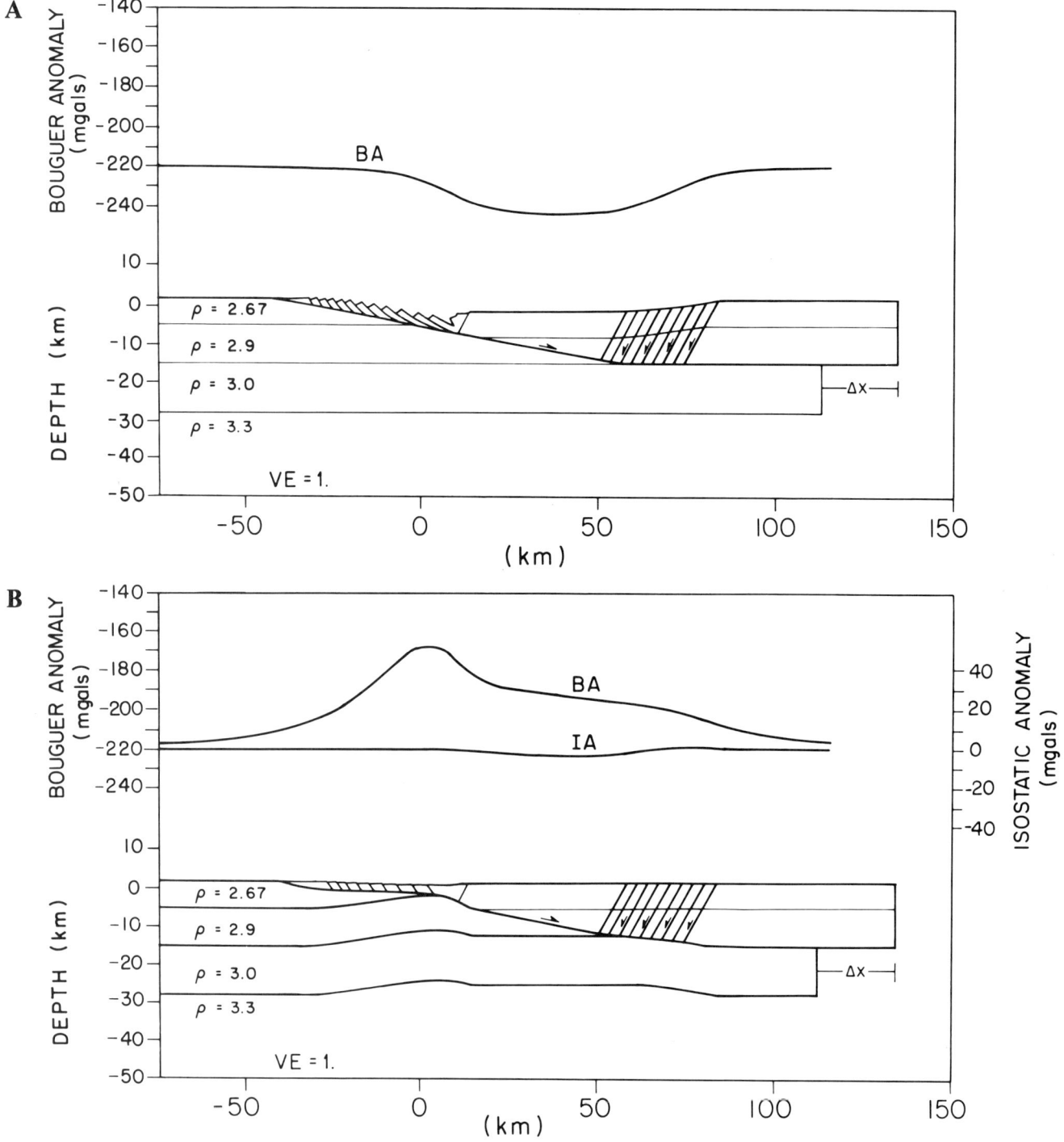

Figure 6. A: Computer-generated model of extension in the upper crust and the Bouguer gravity anomaly before compensation. The regional value of Bouguer anomaly is taken as about −200 mgal to represent usual values in parts of the province that stand about 1.8 km above sea level. If the lower part of the crust were extended by the same amount (Δx) and the extension replaced by intrusions with the density of the lower crust, the Bouguer anomaly would be the same as shown. B: The same model after compensation. Most, but not all, of the topographic altitude has been recovered and the local Bouguer anomaly is relatively positive because of the upwarp of mantle and intermediate crustal layers. The local positive Bouguer anomaly represents isostatic compensation for the decreased altitude of the land. The isostatic anomaly is near zero, reflecting compensation. This model is the same as if the lower crust had been simultaneously extended and the extension replaced by intrusions of the same density as the lower crust.

more silicic components of the crust. These complex processes would strongly influence the patterns of faulting and deformation that we can observe at the surface. A challenge to current and future geophysical exploration of the crust and upper mantle is to provide enough regional coverage and resolution to link the tectonic processes at depth with the observable geologic record.

REFERENCES CITED

Allmendinger, R. W., Sharp, J. W., Von Tish, D., Serpa, L., Brown, L., Kaufman, S., Oliver, J., and Smith, R. B., 1983, Cenozoic and Mesozoic structure of the eastern Basin and Range from COCORP seismic reflection data: Geology, v. 11, p. 532–536.

Anderson, E. R., Zoback, M. L., and Thompson, G. A., 1983, Implications of selected subsurface data on the structural form and evolution of some basins in the northern Basin and Range province, Nevada and Utah: Geological Society of America Bulletin, v. 94, p. 1055–1072.

Angelier, J., and Coletta, B., 1983, Tension fractures and extensional tectonics: Nature, v. 301, p. 49–51.

Bacon, C. R., 1982, Time-predictable bimodal volcanism in the Coso Range, California: Geology, v. 10, p. 65–69.

Barker, F., Wones, D. R., Sharp, W. N., and Desborough, G. A., 1975, The Pikes Peak batholith, Colorado Front Range, and a model for the origin of the gabbro anorthosite-syenite-potassic granite suite: Precambrian Research, v. 2, p. 97–160.

Bott, M.H.P., 1980, Mechanisms of subsidence at passive continental margins, in Bally et al., eds., Dynamics of plate interiors: American Geophysical Union Geodynamic Series, v. 1, p. 27–35.

Brown, L. D., Chapin, C. E., Sanford, A. R., Kaufman, S., and Oliver, J., 1980, Deep structure of the Rio Grande rift from seismic reflection profiling: Journal of Geophysical Research, v. 85, p. 4773–4800.

Cape, C. D., McGeary, S., and Thompson, G. A., 1983, Cenozoic normal faulting and the shallow structure of the Rio Grande rift near Socorro, New Mexico: Geological Society of American Bulletin, v. 94, p. 3–14.

Coney, P. J., 1980, Cordilleran metamorphic core complexes: An overview, in Crittenden, M. D., Coney, P. J., and Davis, G. H., eds., Cordilleran metamorphic core complexes: Geological Society of America Memoir 153, p. 7–34.

Davis, G. A., Anderson, J. L., Frost, E. G., and Shackelford, T. J., 1980, Mylonitization and detachment faulting in the Whipple-Buckskin-Rawhide Mountains terrane, southeastern California and western Arizona, in Crittenden, M. D., Coney, P. J., and Davis, G. H., eds., Cordilleran metamorphic core complexes: Geological Society of America Memoir 153, p. 79–129.

Davis, G. H., 1980, Structural characteristics of metamorphic core complexes, southern Arizona, in Crittenden, M. D., Coney, P. J., and Davis, G. H., eds., Cordilleran metamorphic core complexes: Geological Society of America Memoir 153, p. 35–78.

——, 1983, Shear-zone model for the origin of metamorphic core complexes: Geology, v. 11, p. 342–347.

Eaton, G. P., 1982, The Basin and Range province: Origin and tectonic significance: Annual Review of Earth and Planetary Sciences, v. 10, p. 409–440.

Effimoff, I., and Pinezich, A. R., 1981, Tertiary structural development of selected valleys based on seismic data: Basin and Range province, northeastern Nevada: Royal Society of London Philosophical Transactions, Ser. A, v. 300, p. 435–442.

Evison, F. F., 1960, On the growth of continents by plastic flow under gravity: Royal Astronomical Society Geophysical Journal, v. 3, p. 155–190.

Gans, P. B., and Miller, E. L., 1983, Style of Mid-Tertiary extension in east-central Nevada, in Gurgel, K. D., ed., Guidebook - Part I: Geologic excursions in the overthrust belt and metamorphic core complexes of the intermountain region: Utah Geological and Mineral Survey, Special Studies 59, p. 107–145.

Hamilton, W., 1981, Crustal evolution by arc magmatism: Royal Society of London Philosophical Transactions, ser. A, v. 301, p. 279–291.

Hildreth, W., 1981, Gradients in silicic magma chambers: Implications for lithospheric magmatism: Journal of Geophysical Research, v. 86, p. 10153–10192.

Keller, G. R., Smith, R. B., and Braile, L. W., 1975, Crustal structure along the Great Basin–Colorado Plateau transition from seismic refraction measurements: Journal of Geophysical Research, v. 80, p. 1093–1098.

Lachenbruch, A. H., and Sass, J. H., 1978, Models of an extending lithosphere and heat flow in the Basin and Range province: Geological Society of America Memoir 152, p. 209–350.

McDonald, R. E., 1976, Tertiary tectonics and sedimentary rocks along the transition: Basin and Range province to plateau and thrust belt, Utah: Rocky Mountain Association of Geologists Symposium, p. 281–317.

Miller, E. L., Gans, P. B., and Garing, J., 1983, The Snake Range decollement: An exhumed mid-Tertiary ductile-brittle transition: Tectonics, v. 2, p. 239–265.

Mooney, W. D., Andrews, M. C., Ginzburg, A., Peters, D. A., and Hamilton, R. M., 1983, Crustal structure of the northern Mississippi Embayment and a comparison with other continental rift zones: Tectonophysics, v. 94, p. 327–348.

Moore, J. G., 1960, Curvature of normal faults in the Basin and Range province of the western United States: Article 188 in U.S. Geological Survey Professional Paper 400-B, p. B409–411.

Okaya, D. A., and Thompson, G. A., 1985, Geometry of Cenozoic extensional faulting: Dixie Valley, Nevada: Tectonics, v. 4, p. 107–125.

Proffett, J. M., Jr., 1977, Cenozoic geology of the Yerington district, Nevada, and implications for nature and origin of basin and range faulting: Geological Society of America Bulletin, v. 88, p. 247–266.

Rehrig, W. A., and Reynolds, S. J., 1980, Geologic and geochronologic reconnaissance of a northwest-trending zone of metamorphic core complexes in southern and western Arizona, in Crittenden, M. D., Coney, P. J., and Davis, G. H., eds., Cordilleran metamorphic core complexes: Geological Society of America Memoir 153, p. 131–158.

Richins, W. D., Smith, R. B., Langer, C. J., Zollweg, J. E., King, J. J., and Pechman, J. C., 1985, The Borah Peak, Idaho, earthquake: Relationship of aftershocks to the main shock, surface faulting, and regional tectonics, in Stein, R. S., and Bucknam, R. C., eds., Proceedings, Workshop XXVIII on the Borah Peak, Idaho, earthquake, U.S. Geological Survey Open-File Report 85-290, p. 285–310.

Sanford, A. R., Mott, R. P., Jr., Shuleski, P. J., Rinehart, E. J., Caravella, F. J., Ward, R. M., and Wallace, T. C., 1977, Geophysical evidence for a magma body in the crust in the vicinity of Socorro, N.M.; in Heacock, J. G., ed., The Earth's crust: American Geophysical Union Geophysical Monograph 20, p. 485–503.

Sibson, R. H., 1982, Fault zone models, heat flow, and the depth distribution of earthquakes in the continental crust of the United States: Seismological Society of America Bulletin, v. 72, p. 151–163.

Smith, R. B., and Bruhn, R. L., 1984, Intraplate extensional tectonics of the eastern Basin-Range: Inferences on structural style from seismic reflection data, regional tectonics, and thermal-mechanical models of brittle-ductile deformation: Journal of Geophysical Research, v. 89, p. 5733–5762.

Stein, R. S., and Barrientos, S. E., 1985, The 1983 Borah Peak, Idaho, earthquake: Geodetic evidence for deep rupture on a planar fault, in Stein, R. S., and Bucknam, R. C., eds., Proceedings, Workshop XXVIII on the Borah Peak, Idaho, earthquake: U.S. Geological Survey Open-File Report 85-290, p. 459–484.

Talwani, M., Le Pichon, X., and Ewing, M., 1965, Crustal structure of the mid-ocean ridges; computed model from gravity and seismic refraction data: Journal of Geophysical Research, v. 70, p. 341–452.

Tayor, H. P., Jr., 1980, The effects of assimilation of country rocks by magmas on $^{18}O/^{16}O$ and $^{87}Sr/^{86}Sr$ systematics in igneous rocks: Earth and Planetary Science Letters, v. 47, p. 243–254.

Thompson, G. A., 1959, Gravity measurements between Hazen and Austin, Nevada: A study of basin-range structure: Journal of Geophysical Research, v. 64, p. 217–229.

——, 1960, Problem of late Cenozoic structure of the Basin Ranges: International Geological Congress 21, Copenhagen, Denmark, pt. 18, p. 62–68.

Thompson, G. A., and Burke, D. B., 1974, Regional geophysics of the Basin and Range province: Annual Review of Earth and Planetary Sciences, v. 2, p. 213–238.

Wernicke, B., 1981, Low-angle normal faults in the Basin and Range province: Nappe tectonics in an extending orogen: Nature, v. 291, p. 645–648.

Zoback, M. L., Anderson, R. E., and Thompson, G. A., 1981, Cainozoic evolution of the state of stress and style of tectonics of the Basin and Range province of the western United States: Royal Society of London Philosophical Transactions, ser. A, v. 300, p. 407–434.

Manuscript Accepted by the Society March 4, 1986

ACKNOWLEDGMENTS

Supported by National Science Foundation Grant EAR 81-09294. We owe much to the many colleagues, too many to thank here, who have been building a general body of knowledge about both the geology and the geophysics of the Basin and Range province.

Thermal-mechanical consequences of Basin and Range extension

Kevin P. Furlong
Department of Geosciences
Pennsylvania State University
University Park, Pennsylvania 16802

Michael D. Londe
Department of Geology and Geophysics
University of Wyoming
Laramie, Wyoming 82071

ABSTRACT

Extension zones within continents have complex patterns of tectonic evolution. The Basin and Range Province of western North America provides an ideal location to study the mode of extension in continental regions. We have utilized numerical models to test two distinct geological models of extension that have been proposed for the Basin and Range: (1) a model in which extension takes place by uniform (or pure shear) stretching; and (2) a model in which extension occurs along discrete low-angle shear zones by a simple shear mechanism. These numerical models indicate that both styles of extension produce results generally consistent with observed heat flow, gravity, and elevation data. Distinctive patterns in these data are maintained primarily during the period of extension, implying that present day observations are dominantly a consequence of an ongoing process. The results further imply that the effects of present day extension will obscure the evidence of previous extensional episodes at least as far as the parameters of heat flow, elevation, and gravity are concerned.

INTRODUCTION

The mode of extension within continental regions is considerably more difficult to understand than extension at oceanic spreading centers. Although continental extension zones pass into oceanic rift zones (e.g., East Africa), the specific mechanism for continental extension appears to differ from that of oceanic spreading centers. Attempts to unravel the mode of extension are complicated by the effects of variation in crustal structure (in time and space) and the superposition of the "tectonic fingerprints" of multiple orogenic events.

The Basin and Range Province (B&R) of western North America provides an opportunity to study extension in continental regions. Although the geologic consequences are well studied, the specifics of the extension mechanism within the region are still unresolved. In this study we have investigated, using numerical models, the geophysical responses to two proposed mechanisms of extension. In this way we hope to place realistic constraints on the proposed mechanisms. Specifically, we have modeled (1) a uniform extension mechanism (UEM) for the lithospheric plate (Hamilton and Myers, 1966; Stewart, 1971); and (2) a simple shear mechanism (SSM) where extension occurs along discrete low-angle zones (Wernicke, 1981, 1983, 1985). Previously, UEM has been found to be consistent with heat flow values within the B&R (Lachenbruch and Sass, 1978), while SSM has been proposed as a means of emplacing metamorphic core complexes within the B&R (Wernicke, 1983, 1985) and explaining the geometry of B&R fault zones (Spencer, 1984).

We have evaluated the appropriateness of these models and placed additional constraints on the mode of extension by using surface heat flow data, topography patterns, and the Bouguer gravity data from profiles across the transition zone between the Colorado Plateau (CP) and Basin and Range. These observations were then compared with the geophysical response predicted by our numerical models that simulate UEM and SSM.

The location of the CP to B&R transition zone and the location of the profiles used in constructing the Bouguer gravity and topography profiles are shown in Figure 1. The profiles were

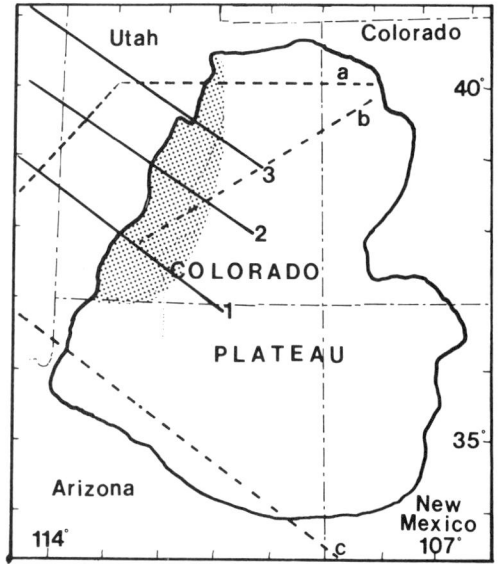

Figure 1. Index map with location of geophysical profiles summarized in Figure 2 (solid lines); labeled dashed lines are locations of seismic refraction lines from Keller and others (1979). Shaded region is approximate location of geophysical transition zone between the B&R and CP.

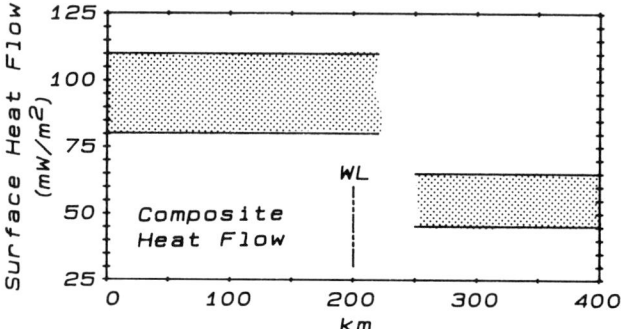
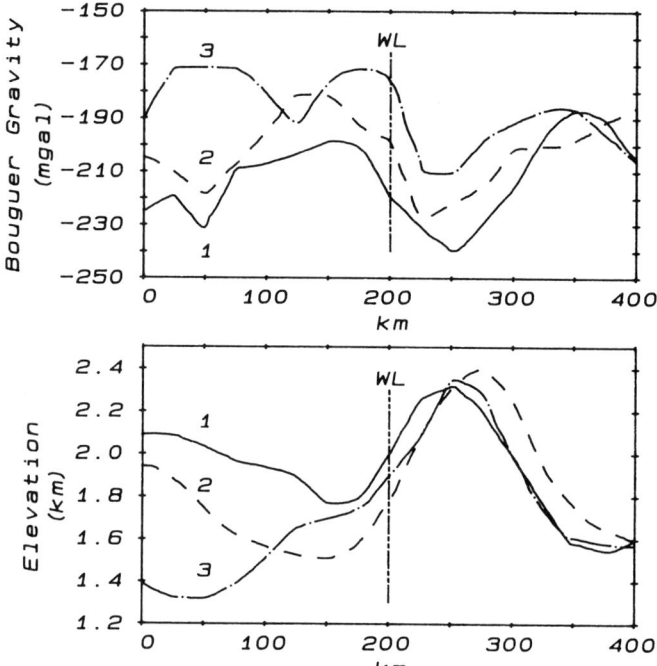

Figure 2. Composite heat flow (Chapman and others, 1978), Bouguer gravity (Woolard and Joesting, 1964), and elevation (Diment and Urban, 1981) profiles across the Basin and Range–Colorado Plateau transition. Gravity and elevation profiles are from locations identified in Figure 1. WL marks the position of the Wasatch Line, the physiographic boundary between the B&R and CP.

aligned subperpendicular to the strike of the transition. Seismic refraction lines (Keller and others, 1979), are also shown, which have been used to constrain crustal thickness in the models.

The general pattern of geophysical data across the transition zone is shown in Figure 2. Heat flow data for this region have been summarized in a number of sources (Chapman and others, 1978; Blackwell, 1978; Keller and others, 1979; Thompson and Zoback, 1979; Lachenbruch and Sass, 1977, 1978; Reiter and Clarkson, 1983; and Bodell and Chapman, 1982). The shaded bands indicate the range of values in each heat flow province. Heat flow values within the B&R are high (85–110 mW m^{-2}) while within the interior of the CP values are much lower (45–66 mW m^{-2}). The nature of the transition in the heat flow data between the two regions is not well constrained.

The Bouguer gravity anomalies and topography were digitized at a 25-km interval from Woolard and Joesting (1964) and Diment and Urban (1981), respectively. There is a general decrease in the Bouguer value across the transition zone of 30–50 mgals, with a gravity low corresponding to the location of the topographic highs. This step is most clearly seen in the two northern profiles (cf. Fig. 3, Thompson and Zoback, 1979).

The high plateaus of the transition zone stand approximately 500 to 1000 m higher than either the B&R or the CP interiors. Within the region the crustal thickness is apparently increasing from 25–30 km within the B&R to 40–50 km beneath the CP interior, although variation in crustal thickness is poorly understood.

In this study, we have intentionally used simplified geological and geophysical assumptions as constraints in our models. We feel that by first studying the results of these simple models and understanding their consequences, we will be able to construct more refined models that will provide further insight into the physical processes active in the region.

MODELING APPROACH

The modeling involved both one-dimensional (1-D) and two-dimensional (2-D) numerical models of the thermal response to extension. The finite difference method was used to solve the appropriate form of the heat transfer equation within a region chosen to simulate the extending zone, transition, and stable block. The geometry used for testing the UEM model is shown in

Figure 3. The region modeled extends from the interior of the B&R to the interior of the CP (total distance of 400 km). In this way we could be sure that the thermal behavior in the transition zone was not affected by our choice of boundary conditions at the right and left boundaries of the model. The 2-D geometry was used to test the role of lateral heat transport and the resulting gravity and topography signatures of the lithospheric flexure caused by variations in crustal structure and thermal state. As shown in Figure 3, two different crustal structures were used in the modeling, which differ in the location of the step change in crustal thickness.

The SSM model was tested using a series of 1-D (vertical) time-dependent models that simulated the vertical mass and heat transfer during extension, and the thermal relaxation after cessation of extension. Because of the proposed low-angle nature of the shear zone, the primary disturbances to heat transfer are from the vertical direction, since horizontally adjacent regions have comparable temperatures, and thus little lateral heat transfer. The series of 1-D models were then linked together to provide an estimate of the two-dimensional thermal structure and topographic response. The elevation profiles were determined by considering the effects of variations in the thermal regime as well as changes in crustal structure that result from SSM extension.

UNIFORM EXTENSION MODEL (UEM)

Previous work (Lachenbruch and Sass, 1978) has demonstrated that UEM can explain elevated heat flow observed in the interior of the B&R, and the general lack of mid-crustal melt zones. However, 1-D models are appropriate only to the interior of the extending region, far from the boundaries, and do not allow investigation of the nature of the transition from the extending region to the stable block (B&R to CP). The use of 2-D modeling for the extension model allows investigation of the geophysical behavior in this transitional region.

In this modeling, the steady-state heat transfer equation with mass transfer and heat source terms:

$$\mathbf{u} \cdot \nabla T = K \nabla^2 T + (A_o/\rho c)\exp(-z/D) \quad (1)$$

was solved, where $\mathbf{u}$ is the velocity vector (two dimensional), K is thermal diffusivity, ρ is density, c is specific heat, and A_o is surface heat production. Heat production is assumed to have an exponential decrease with depth, i.e., $\exp(-z/D)$, with D as the characteristic depth for this decrease (Lachenbruch, 1970). Equation 1 was solved on a 31 × 11 finite difference grid with physical dimensions of 400 km by 65 km (Fig. 3). Grid spacings of 13.3 km horizontal and 6.5 km vertical were used. Additional parameters used in the UEM models are $A_o = 2.5\ \mu W\ m^{-3}$, $K = 10^{-6}\ m^2$ s, and $D = 10$ km. The horizontal extent of the model spans the distance from the B&R interior to the CP interior, and the vertical extent (65 km) represents the lithospheric section in the B&R, consistent with surface wave studies of Priestley and Brune (1978). This domain was partitioned into two regions: (1) the

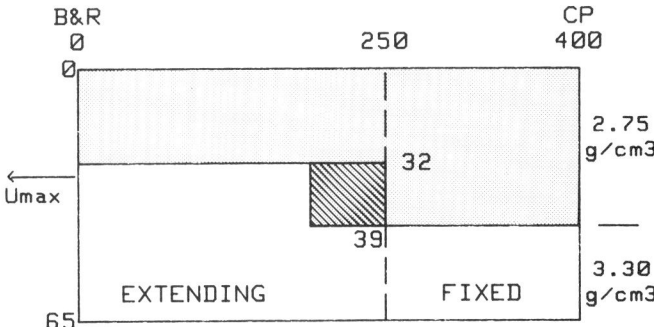

Figure 3. Schematic cross section of UEM model domain and boundary conditions. All dimensions are in kilometers. Shaded area represents crustal section of model. Hatched and shaded area is additional section of crust in alternate crustal structure used in modeling (see text for details).

eastern or righthand part of the grid represents the western section of the CP where heat is transported purely by conduction and there is no extension (i.e., $|\mathbf{u}| = 0$), and (2) the western or lefthand part of the model grid represents the B&R and B&R-CP transition, where there is mass transfer (i.e., $|\mathbf{u}| > 0$) and heat is transferred both by conduction and advection. The boundary between the extending and stable blocks was placed 150 km from the eastern side of the model (distance label 250 km), while the assumed position of the Wasatch Line was arbitrarily placed 75 km west of the edge of the stable block representing the CP interior.

The velocity field, $\mathbf{u}$, for the stretching lithosphere was computed by specifying the maximum horizontal velocity (at the lefthand boundary). This allowed the horizontal strain rate, and hence horizontal and vertical velocity fields to be computed, given UEM assumptions. Maximum horizontal velocities of 0.25 cm/yr, 0.5 cm/yr, and 1.0 cm/yr were used in the calculations. These velocities correspond to strain rates of 1%/m.y., 2%/m.y., and 4%/m.y., respectively.

The boundary conditions used to solve equation 1 were chosen from limits on observed heat flow and strain rates. The top surface was maintained at 0°C. Temperatures at the righthand boundary (interior of the CP) were given by a geotherm appropriate for a surface flux of 65 mW m^{-2} with a crustal (radiogenic) contribution of 25 mW m^{-2}. Mixed conditions were applied to the bottom boundary. A basal flux condition of 40 mW m^{-2} was applied in the stable region (CP), while the base of the extending region (B&R) was kept at a temperature of 1250°C (simulating asthenospheric conditions). Temperatures at the lefthand boundary (interior of B&R) were given by a one-dimensional form of the conduction/advection equation equivalent to the geotherm given by Lachenbruch and Sass (1978) for crystalline stretching.

Results for the temperature structure and surface heat flow are given in Figure 4. The overall patterns of temperature and heat flow are similar for the three extension rates, but the greater strain rates, with associated more rapid vertical advection of heat,

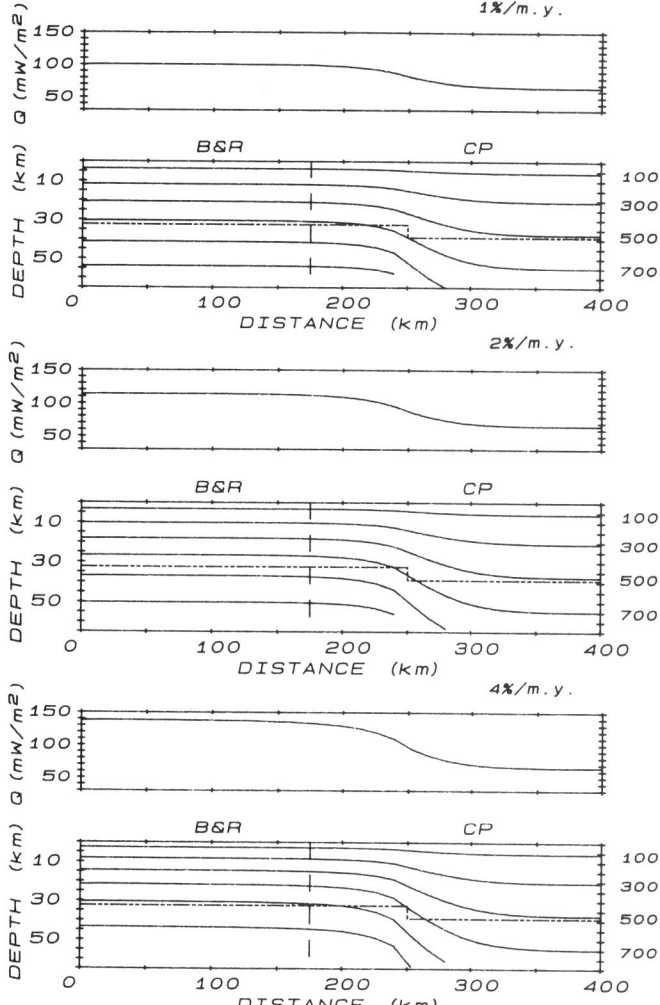

Figure 4. Results of thermal modeling of UEM. Surface heat flow profiles and contoured temperature fields for assumed strain rates of 1%/m.y., 2%/m.y., and 4%/m.y. are given. Dot-dash line represents base of crust. Temperature contours are labeled and are every 200°C (beginning at 100°C. Vertical dashed line is inferred position of Wasatch Line.

cause a shallowing of the isotherms. Within the fixed block temperatures within 30 to 50 km of the mechanical boundary are elevated relative to the CP interior as a consequence of lateral heat conduction from the extending (B&R) into the stable (CP) regions. Between the mechanical boundary and the actively extending region, there is a transition zone 50 to 75 km in width of increasing heat flow (and increasing crustal temperatures) beyond which the thermal regime is essentially laterally uniform. Consequently, these models predict an overall thermal transition 80 to 125 km in width approximately centered on the mechanical boundary between the fixed and extending blocks.

The uplift response was computed for the two crustal models shown in Figure 3, assuming elastic flexural behavior for the lithosphere. The applied load driving the flexure was determined by calculating the excess mass in each vertical column (vertical array of grid points) relative to a reference column (rightside boundary). This load was assumed to act as a line load on the lithosphere (extending in the third, or north-south, direction). The total elevation response was determined as a superposition of the responses from each of the line loads. Flexural rigidities of 10^{21}, 10^{22}, and 10^{23} Nm (Newton-meters) were used in these calculations. The entire lithosphere was assumed to act as an elastic plate with an effective thickness determined from the best-fitting flexural rigidities. This avoids the need to *a priori* partition the lithosphere into elastic and non-elastic regions.

The results of the uplift modeling are given in Figures 5 and 6. When the crustal step is placed at the mechanical boundary (Fig. 5), heating of the fixed block by lateral conduction causes increases in elevation of 100 to 200 m, while within the extending block there is 300 to 500 m of subsidence. The major elevation gradients are confined to a zone 100–150 km wide, roughly centered on the mechanical boundary. If the thicker crust extends west of the mechanical boundary (Fig. 6), the major elevation increase is in the region of thick crust *and* extension. Elevation increases of as much as 500 m occur in this zone. Thus the elevation response to extension in the B&R is sensitive to both the rates of extension and the crustal structure in the transition zone.

The calculated effects of the variations in the temperature field and crustal structure on gravity are shown in Figure 7. The surface gravity field is calculated using a sequence of line masses (extending into the third dimension of the region, i.e., approximately N-S) located at each node point, representing the mass anomalies in each cell. Again the results are presented for two crustal structure models. When the crustal thickness varies at the mechanical boundary, there is a calculated gravity step of 30 to 60 mgal. When the crustal step is within the extension region, there is again a 30–60 mgal step across the transition zone, but now there is also a local gravity low of 10–30 mgal within the region of maximum uplift, consistent with the observed gravity field.

The results of this simple modeling demonstrate first order agreement between the observed and calculated geophysical signatures across the transition (Fig. 2). The observed surface heat flow is well matched by predicted heat flow (Fig. 4) for extension rates of 1%–2%/m.y. The elevation patterns are consistent with the uplift response predicted by the model when flexural rigidities are in the range of 10^{21} to 10^{22} Nm. Additionally, both the gravity and topography data are most consistent with the crustal model of a thicker crust extending beyond the stable region approximately 40 km into the extending region. This crustal structure, of thick crust under the high plateaus of east-central Utah, argues that a significant variation in crustal structure occurs at the Wasatch Line (Smith, 1978). These results lend support to a model of lithospheric extension in the Basin and Range, which, on the scale sampled by these geophysical parameters, is equivalent to UEM. The geophysical data observed in the B&R-CP transition zone are likely a consequence of edge effects of the thermal-mechanical features of the extending zone.

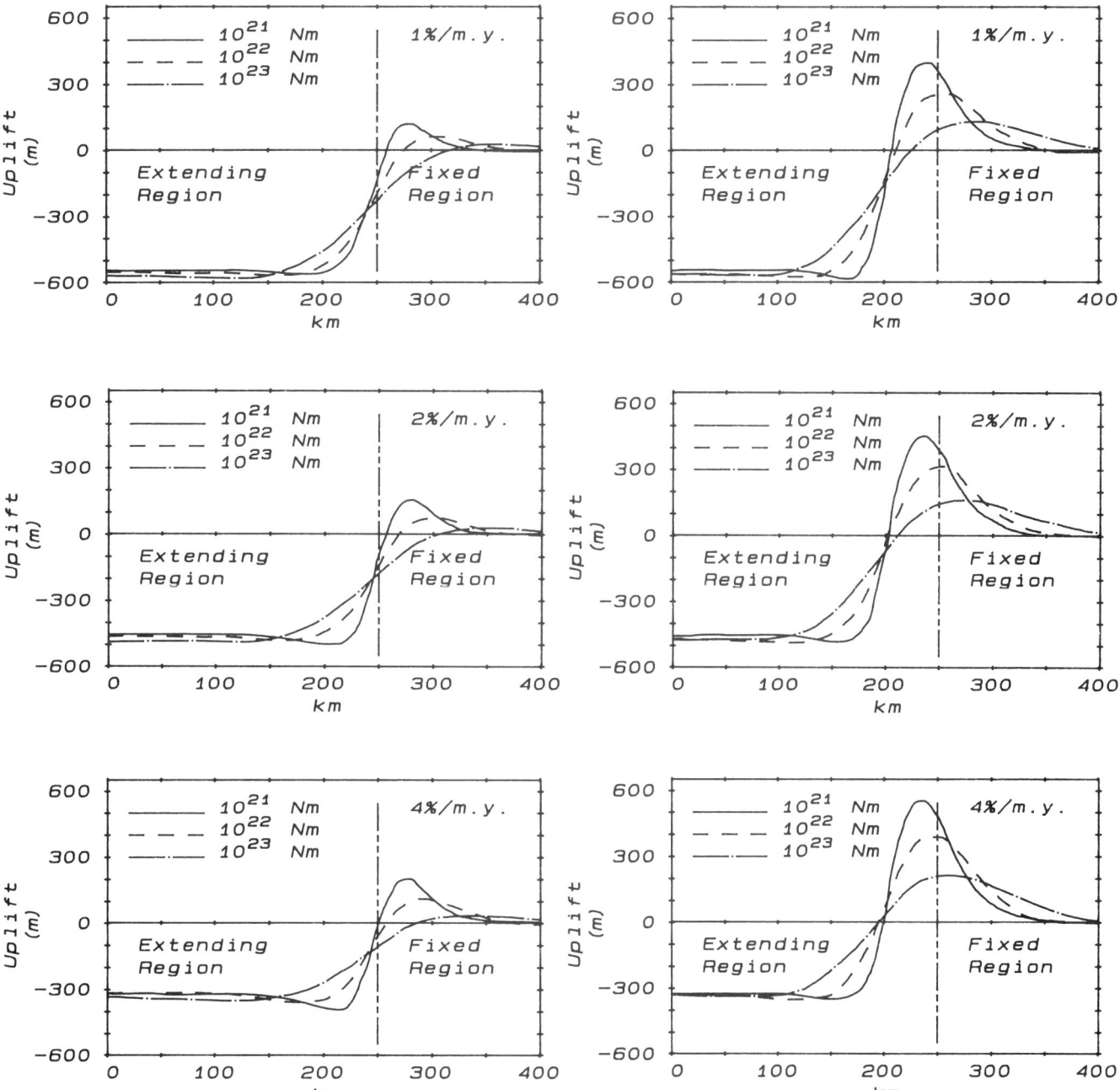

Figure 5. Predicted uplift patterns resulting from crustal thermal expansion and plate flexure. Extension rates of 1%/m.y., 2%/m.y., and 4%/m.y., and model results of Figure 4 were used in these calculations. Flexure models are based on flexural rigidities of 10^{21}, 10^{22}, and 10^{23} Nm (Newton-meters; standard crustal model).

Figure 6. Predicted uplift patterns from use of alternate crustal model (see Fig. 3 and text). Other conditions as given in Figure 5.

SIMPLE SHEAR EXTENSION MODEL (SSM)

In a series of recent papers, Wernicke (1981, 1983, 1985) has proposed that a significant amount of the total crustal extension in the B&R occurs by simple shear along low-angle shear zones, which cut the entire crust and perhaps extend into or through the lithospheric mantle. Wernicke has argued that this mechanism explains observed patterns of metamorphic grade and patterns of ductile and brittle deformation within the B&R. Here, using numerical models, we have tested the geophysical signatures (heat flow, elevation) of this model of simple shear extension as a means of further evaluating the appropriateness of SSM.

In modeling SSM, we have used a series of one-dimensional (1-D) time-dependent numerical models (finite difference) that are linked together to provide an estimate of the 2-D heat flow

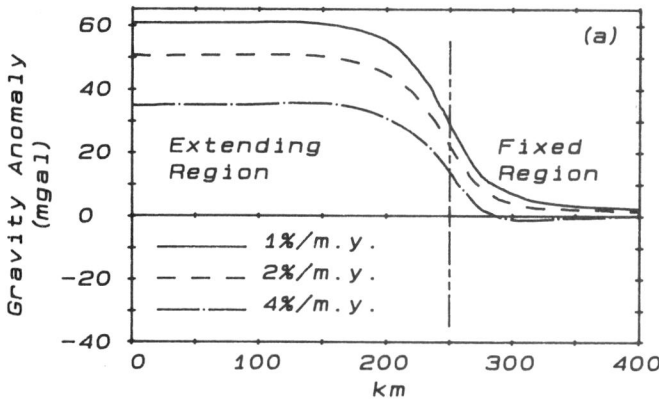

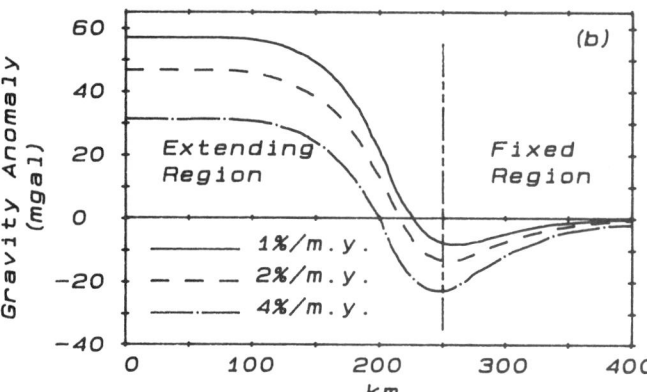

Figure 7. Computed gravity anomaly for UEM models and extension rates of 1%/m.y., 2%/m.y., and 4%/m.y. (a) standard crustal model, (b) alternate crustal model.

TABLE 1. PARAMETERS FOR SSM MODEL

Parameter	Value
Spatial Discretization	0.5 km
Time Discretization	0.125 Ma
Thermal Diffusivity	10^{-6} m^2 s^{-1}
Thermal Conductivity	3.0 W m^{-1} K^{-1}
Surface Heat Production (Ao)	2.5 uW m^{-3}
Heat Production Decrement[1]	10 km
Surface Temperature	0°C
Asthenospheric Temperature	1300°C
Lithospheric Thickness (initial)	90 km
Crustal Thickness (initial)	45 km
Surface Heat Flow (initial)	62.5 mW m^{-2}
Crustal Density[2]	2700 kg m^{-3}
Mantle Density[2]	3400 kg m^{-3}
Thermal Expansion	3 x 10^{-5} °C^{-1}

[1]Heat production assumed to follow exponential decrease with depth (Lachenbruch, 1970).
[2]Density at 0°C.

and elevation patterns produced by SSM. The 1-D approach provides reasonable estimates of the thermal response because the assumed low-angle nature of this shear zone (ca. 20° dip) results in primary disturbances to the thermal field from vertical heat transport. The effects of extension are simulated by the upward transport of mass and associated heat content with concurrent heat conduction. An important benefit of using the "linked 1-D" model is that the results are independent of the specific geometry of the shear zone, allowing relatively general application of the model. The 1-D assumptions have been previously used successfully in regions of low-angle simple shear to model the thermal evolution of thrust regimes (Oxburgh and Turcotte, 1974; Brewer, 1981; Angevine and Turcotte, 1983; Furlong and Edman, 1984).

To generate the results shown here, we have assumed that extension occurred over a time period of 15 m.y. with total vertical throw of 20 km. With a shear zone dipping at 20°, this corresponds to a horizontal extension velocity of 3.6 mm/yr. Additional parameters used in the modeling are given in Table 1.

A schematic of the modeled region is shown in Figure 8. This cross section extends from the B&R interior across the transition zone into the interior of the CP. This diagram represents a hypothetical crustal cross section after simple shear extension has ended (points x and x' were initially adjacent). The model extended to a depth of 100 km with an initial lithospheric thickness of approximately 90 km. The horizontal extent of the region shown in Figure 8 depends on the assumed dip of the shear zone. With a dip of 20°, the region is approximately 280 km wide. Using the terminology of Wernicke (1985), the surface location labeled A corresponds to the location of extensional allochthons (core complexes), B corresponds to the limit of significant upper crustal extension, and C is the location of the Moho "hinge." With a 20° dip, the region spanned from A to C is approximately 140 km wide. The "discrepant zone" of Wernicke (1985) is the region between B and C.

Lithospheric thickness across this region will likely vary in space and time both during and after extension. The results of our modeling indicate that for the extension rates used, as a consequence of the rate at which the "asthenospheric" mantle cools to become "lithospheric" mantle, net lithospheric thinning during the period of extension is approximately one half of the magnitude of crustal thinning (<10 versus 20 km). For higher extension rates, this net transient lithospheric thinning will be greater.

For the SSM model we determined the thermal structure through time and from the thermal results computed the predicted surface heat flow and elevation response for the region between locations A and C, the region containing the transition zone from the B&R to the CP. The region to the left (west) of A in general would have a heat flow and elevation history similar to that at A with disturbances decreasing as one moves further to the left (west). The duration behavior in this region is difficult to predict as that response would be very sensitive to assumptions of erosion. The region to the right (east) of C would be only slightly affected by the extensional event.

The predicted surface heat flow profiles are given in Figure 9a, during extension (5 m.y.), immediately after extension (15

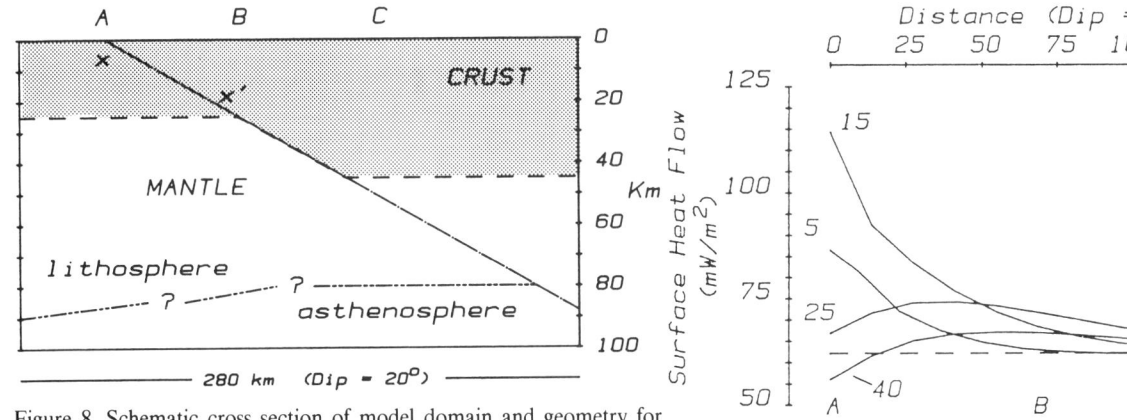

Figure 8. Schematic cross section of model domain and geometry for SSM model. Shaded region represents crust. A, B, and C mark locations referred to in text. Positions labeled x and x' were adjacent prior to extension.

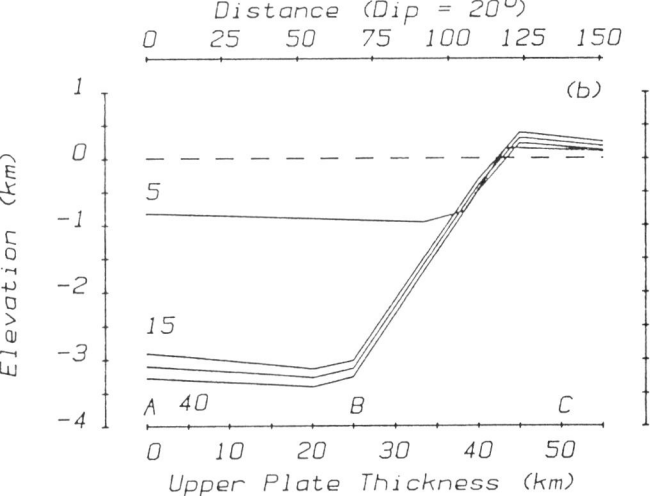

Figure 9. (a) Predicted surface heat flow response during (0–15 m.y.) and after (>15 m.y.) SSM extension. Dashed line is initial heat flow regime. (b) Predicted elevation response during and after SSM extension. Dashed line is initial elevation value. Numbers labeling curves are time in m.y. since start of extension.

m.y.), and at two later times (25 m.y., 40 m.y.). Locations are labeled both in terms of the thickness of the upper plate and the distance from position A, assuming a 20° dip for the shear zone. The heat flow response can be summarized as follows. During and immediately after extension the maximum heat flow anomaly occurs in the region of thinnest upper plate. After extension ends this region cools and the heat flow on this area decreases reaching an equilibrium value lower than the initial value because of the removal of (assumed) radiogenically enriched uppermost crust. The regions with a thicker upper plate show a time lag in the surface heat flow response with a progressive rightward (eastward) shift in the location of heat flow anomaly maximum. The width of the heat flow transition varies in time, but has a typical width (assuming 20° dip) of 75 km. This compares favorably with the width of the observed heat flow transition. The modeling indicates that the heat flow transition should occur in the region between the core complexes and the end of the zone of near-surface extension; the observed heat flow transition is more to the east of that zone in the region of the high plateaus.

The elevation response was calculated assuming local compensation (required by our 1-D assumptions) incorporating both the effects of temperature and crustal structure variations. The elevation response is dominated by the effects of variations in crustal structure. *Average* temperatures in the lithospheric section vary by <200°C restricting effects of density variations from temperature changes on elevation to <600 m. Crustal thickness varies by as much as 20 km, leading to an elevation variation of >3 km. Smaller amounts of total extension will decrease this figure but in all cases the crustal structure effect will dominate.

The predicted elevation profiles are shown in Figure 9b. More than 85% of the elevation variation occurs during the period of extension. There is rapid subsidence in the region between A and B, with a transition zone in the elevation response between B and C. In location C, there is very little variation in elevation from this mechanism. The overall elevation response is similar to that observed, but as in the UEM modeling, the equivalency of elevations in the CP interior and the B&R is not produced.

DISCUSSION

The modeling results indicate that the observed pattern of decreasing heat flow from values typical of the B&R across a transition to lower values typical of the CP requires active extension. The UEM model with its steady-state (dynamic) assumptions generates such a pattern. SSM also produces a similar pattern but only during extension and for perhaps 15 m.y. after extension ceases. It appears that a transition in heat flow would be difficult to maintain in any static regime. The elevation patterns produced by both models are roughly consistent with the observed topography, but both models fall short in matching the observed elevation profile. In both cases the elevation effect is very sensitive to the assumed crustal structure, and we are unable to match elevation patterns with either model using current estimates of the crustal structure.

The SSM leaves its mark primarily in crustal structure and hence elevation. Any pre-Miocene extension by this mechanism in the B&R would not have any significant effect on the present day thermal field, and can be delineated with the parameters tested here only insofar as elevation and crustal thickness patterns in the past can be determined. Additionally, if extension in the B&R occurs with some combination of mechanisms, the similar patterns of heat flow and elevation produced by these two models indicates that observations other than geophysical ones are necessary for resolving the relative importance of the two styles of extension.

CONCLUSION

From the results of the relatively simple numerical models presented here, two important points become clear. First, the general pattern of present day geophysical observations can be reasonably well matched by the geophysical response of the crust to the uniform extension model. Second, the geophysical response to SSM is quite significant during the period of extension, but by 15 m.y. after SSM extension ceases the modeling can not be constrained by observed heat flow or uplift/subsidence patterns. In order to identify occurrences of SSM in the past, paleo-indicators of temperature and elevation are needed. If UEM is a currently active extension mechanism, the investigation of SSM becomes all the more difficult because of the significant obscuring effect of the tectonic overprinting from UEM.

ACKNOWLEDGMENTS

This work was partially supported by NSF grant number EAR-8219113 (to KPF) and an AMOCO Ph.D. Fellowship (to MDL). Discussion of this work with D. S. Chapman at the University of Utah was instrumental in its development. Detailed reviews by M. L. Zoback and D. D. Blackwell helped improve this paper and are appreciated.

REFERENCES CITED

Angevine, C. L., and Turcotte, D. L., 1983, Oil generation in overthrust belts: American Association of Petroleum Geologists Bulletin, v. 67, p. 235–241.

Blackwell, D. D., 1978, Heat flow and energy loss in the western United States, *in* Smith, R. B., and Eaton, G. P., eds., Cenozoic tectonics and regional geophysics of the western Cordillera: Geological Society of America Memoir 152, p. 175–208.

Bodell, J. M., and Chapman, D. S., 1982, Heat flow in the north central Colorado Plateau: Journal of Geophysical Research, v. 87, p. 2869–2884.

Braile, L. W., Keller, G. R., and Peeples, W. J., 1974, Inversion of gravity data for two-dimensional density distributions: Journal of Geophysical Research, v. 79, p. 2017–2021.

Brewer, J., 1981, Thermal effects of thrust faulting: Earth and Planetary Science Letters, v. 56, p. 233–244.

Chapman, D. S., Furlong, K. P., Smith, R. B., and Wechsler, D. J., 1978, Geophysical characteristics of the Colorado Plateau and its transition to the Basin and Range Province in Utah, paper presented at The Conference on Plateau Uplift: Flagstaff, Arizona, Mode and Mechanism, Lunar and Planetary Science Institute Topical Conference, p. 14–16.

Diment, W. H., and Urban, T. C., 1981, Average elevation map of the conterminous United States (Gilluly averaging method): U.S. Geological Survey Geophysical Investigations Map GP 933, scale 1:2,500,000.

Furlong, K. P., and Edman, J. D., 1984, Graphic approach to determination of hydrocarbon maturation in overthrust terrains: American Association of Petroleum Geologists Bulletin, v. 68, p. 1818–1824.

Hamilton, W., and Myers, W. B., 1966, Cenozoic tectonics of the western United States: Reviews of Geophysics and Space Physics, v. 5, p. 509–549.

Keller, G. R., Braille, L. W., and Morgan, P., 1979, Crustal structure, geophysical models, and contemporary tectonics of the Colorado Plateau: Tectonophysics, v. 61, p. 131–147.

Lachenbruch, A. H., 1970, Crustal temperature and heat production: Implications of the linear heat-flow relation: Journal of Geophysical Research, v. 75, p. 3291–3300.

Lachenbruch, A. H., and Sass, J. H., 1977, Heat flow in the United States and the thermal regime of the crust, *in* Heacock, J. G., ed., The Earth's crust: American Geophysical Union Geophysical Monolog 20, p. 626–675.

—— , 1978, Models of an extending lithosphere and heat flow in the Basin and Range Province, *in* Smith, R. B., and Eaton, G. P., eds., Cenozoic tectonics and regional geophysics of the western Cordillera: Geological Society of America Memoir 152, p. 209–250.

Oxburgh, E. R., and Turcotte, D. L., 1974, Thermal gradients and regional metamorphism in overthrust terrains with special reference to the eastern Alps: Schweizerische Mineralogische und Petrographische Mitteilungen, v. 54, p. 641–662.

Priestley, K., and Brune, J., 1978, Surface waves and the structure of the Great Basin of Nevada and western Utah: Journal of Geophysical Research, v. 83, p. 2265–2272.

Reiter, M., and Clarkson, G., 1983, A note on terrestrial heat flow in the Colorado Plateau: Geophysical Research Letters, v. 10, p. 929–932.

Smith, R. B., 1978, Crustal structure, and intraplate tectonics of the Wyoming-Idaho-Utah thrust belt, *in* Smith, R. B., and Eaton, G. P., eds., Cenozoic tectonics and regional geophysics of the western Cordillera: Geological Society of America Memoir 152, p. 111–144.

Spencer, J. E., 1984, The role of tectonic denudation in the warping and uplift of low angle normal faults: Geology, v. 12, p. 95–98.

Stewart, J. H., 1971, Basin and Range structure: A system of horsts and grabens produced by deep-seated extension: Geological Society of America Bulletin, v. 82, p. 1019–1044.

Thompson, G. A., and Zoback, M. L., 1979, Regional geophysics of the Colorado Plateau: Tectonophysics, v. 61, p. 149–181.

Wernicke, B., 1981, Low-angle normal faults in the Basin and Range Province: Nappe tectonics in an extending orogen: Nature, v. 291, p. 645–648.

—— , 1983, A simple relationship between extensional belts and plateau uplift: EOS Transactions, American Geophysical Union, v. 64, p. 856.

—— , 1985, Uniform-sense normal simple shear of the continental lithosphere: Canadian Journal of Earth Science, v. 22, p. 108–125.

Woolard, G. P., and Joesting, H. R., 1964, Bouguer gravity anomaly map of the United States: Reston, Virginia, U.S. Geological Survey, scale 1:2,500,000.

MANUSCRIPT ACCEPTED BY THE SOCIETY MARCH 4, 1986

Tertiary structural development of selected basins: Basin and Range Province, northeastern Nevada

I. Effimoff
Ashland Exploration, Inc.
900 Threadneedle, Suite 800
Houston, Texas 77079-2907

A. R. Pinezich
Shell Oil Company
Houston, Texas 77001

ABSTRACT

Reflection seismic data in the Railroad, Diamond, Mary's River, and Goshute valleys provide information on their structural development that cannot be deduced solely from outcrop and well data.

These valleys contain Tertiary sediments that, in dip section, define asymmetrical basins each bounded along the eastern flank by a major listric normal fault with about 3.0–4.6 km (10,000–15,000 ft) of displacement. The western flanks are defined by gentle east-dipping ramps. Seismically, the surfaces of the listric faults are interpreted to dip westward and become bedding-parallel within the Paleozoic sequence, perhaps exploiting regionally recognized Mesozoic decollement surfaces. The Tertiary depocenters, adjacent to the faults, shifted from west to east with continued slippage through time, the greatest movement occurring in Miocene and post-Miocene time. In the strike direction, the basins are separated into at least two sub-basins by an east-west, structurally high axis. The axes are postulated to be the result of a tear fault associated with movement along the listric normal fault.

The Tertiary stratigraphy varies between basins and between sub-basins in a given valley. All the basins contain Miocene and younger rocks; however, not all sub-basins contain the pre-Miocene sequence, suggesting a complex scheme of structural development.

INTRODUCTION

The purpose of this study is to demonstrate the similarities and differences in the Tertiary structural development of selected basins in the Basin and Range Province of northeastern Nevada (Fig. 1). Railroad, Diamond, Mary's River, and Goshute valleys were chosen as examples for this study because of the availability of abundant geophysical and geological data. Integration of common depth point reflection seismic data with outcrop and well information provides documentation on the structural development of the valleys.

A number of recent publications are devoted to the geological-geophysical framework and the regional tectonic significance of the Basin and Range Province of the western United States (Newman and Goode, 1979; Smith and Eaton, 1978; Armentrout and others, 1979; Miller and others, 1983; Zoback and others, 1981; particularly comprehensive overviews are provided by Eaton, 1979, 1982, and Stewart, 1978). There is general agreement that the present physiographic configuration of the province is the result of Cenozoic extensional deformation and that the precise timing of this deformation may be different for different sections of the province.

Two general models for basin and range structural development have been proposed in the literature and are illustrated in Figure 2, which is modified after Stewart (1978). The first model relates to curved, downward-flattening normal (listric) faults

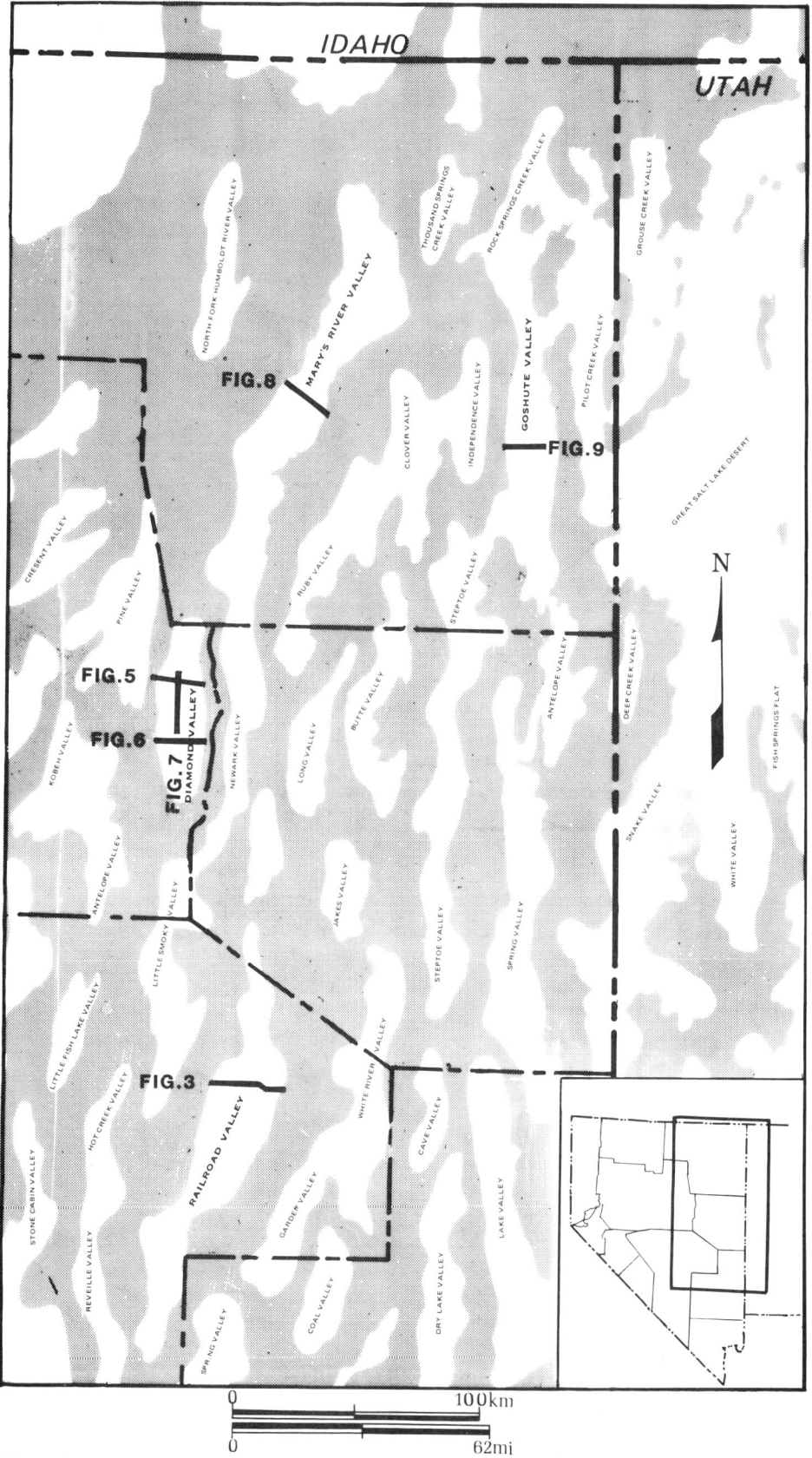

Figure 1. Index map for northeastern Nevada illustrating distribution of basins and ranges. Seismic profiles discussed in text are indicated by figure number.

where the uptilted part of a block forms the range and the down-tilted part forms the valley. The second model consists of horsts and grabens that form ranges and valleys, respectively. The structures observed in the four valleys chosen for this study favor the first model over the second. However, certain minor elements of the second model can be observed. COCORP seismic reflection data collected from the eastern Basin and Range also support the first model (Allmendinger and others, 1983). Other models for the Basin and Range province have been proposed, for example, by Anderson and others (1983); who combine the two models.

Listric faults in other geologic provinces have been identified using reflection seismic data. Of particular note are papers by Bally and others (1966), Lowell and others (1975), Montadert and others (1979), and Bally (1981).

RAILROAD VALLEY

Railroad Valley (Figs. 3 and 4) is geologically one of the most extensively studied and best understood basins in Nevada due to the petroleum industry's activity in the valley for more than 30 years. Numerous wells have been drilled, and many seismic and other geophysical surveys have been conducted in the exploration/production process. Currently Railroad Valley is one of two basins in Nevada with oil production. Eagle Springs and Trap Springs fields are well described by Bortz and Murray (1979) and Duey (1979). Other production in the valley is reported by Petroleum Information Corporation (1984).

The interpretation of the seismic dip profile for Railroad Valley presented in Figure 3 is the result of integrating pertinent well bore and outcrop data with the seismic data. The age units indicated were extrapolated on the profile on the basis of the seismic character of the reflection packages. Gravity data were also integrated into the interpretation. The seismic profile is representative of Railroad Valley in terms of its structural geometry and stratigraphy. The generalized Tertiary stratigraphy of Railroad Valley is shown on Figure 4.

The Tertiary sediments in Railroad Valley define an asymmetrical basin bound on the east by a listric fault along which the Paleozoic is displaced 4.5 km (15,000 ft) vertically. The fault plane is marked by the cessation of reflectors. The listric fault passed downdip into a more prominent west-dipping, low-angle plane; the westward extent is obscured by data quality and the fact that data were recorded to 3.0 seconds two-way time. The fault therefore can be traced only to about the middle of the profile. To the east, the low-angle listric fault appears to pass into a decollement surface. This surface continues beneath the Grant Range flanking the basin. The decollement surface has not been observed in the outcrop of the Grant Range (Kirkpatrick, 1960), but it can be postulated that it is the Mississippian Chainman Shale, a regionally recognized decollement surface. The Chainman has been observed in a possible foot wall position of the basin during the development drilling at Eagle Springs field (Bortz and Murray, 1979).

The western flank of the Tertiary basin is defined by a gently

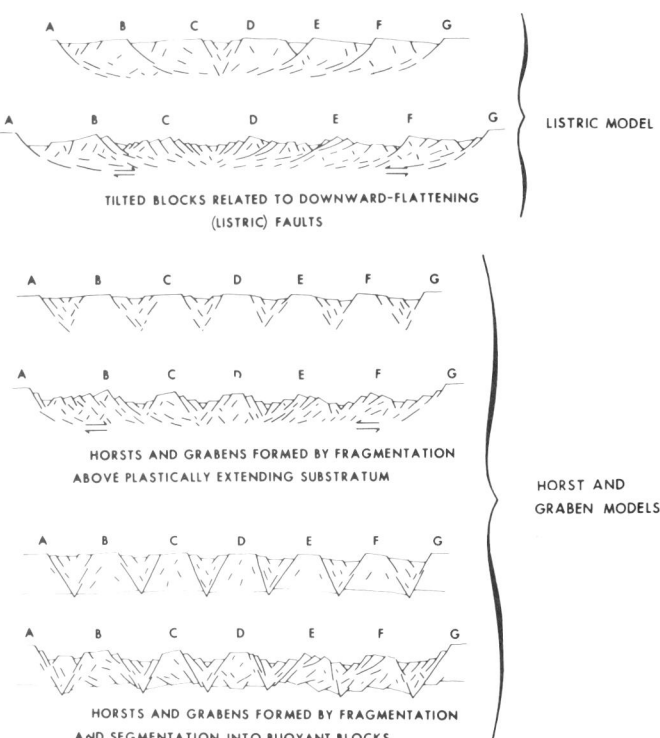

Figure 2. Two general models for basin and range structural development showing early and late stages of development.

eastward-dipping ramp and is in a hanging wall relationship to the low-angle, westward-dipping listric fault. In the hanging wall, Paleozoic units extend across approximately the western two-thirds of the basin and lower Tertiary units extend over the remaining eastern third. The foot wall consists of Paleozoic units exclusively. The geometric relationship of the Tertiary units to the listric fault suggest that the basin may have been initiated by extension in early Tertiary time as suggested by the thinning of the lower Tertiary section toward the west flank of the basin. This observation is supported by detailed stratigraphic studies of the Upper Cretaceous (?) to lower Oligocene Sheep Pass Formation which directly overlies the Paleozoic strata (Bortz and Murray, 1979; Fouch, 1977, 1979; Winfrey, 1960; Newman, 1979), and it can also be inferred from regional studies of Tertiary lacustrine sediments (McDonald, 1976).

Basin development progressed at a more accelerated rate during late Tertiary (Miocene and post-Miocene) time, as evidenced by the thickness of this unit, which comprises the bulk of the Tertiary sequence in the basin. The depocenter of each seismically discernible upper Tertiary unit migrates from west to east as slippage occurs along the low-angle listric fault. Some minor normal faulting effecting lower units of the upper Tertiary, the lower Tertiary, and Paleozoic units in the hanging wall is interpreted to have occurred in early late Tertiary time. The minor faulting in the eastern half of the basin is observed to pass into the low-angle listric fault. In the western half of the basin the minor faults cannot be directly traced into the listric fault because of

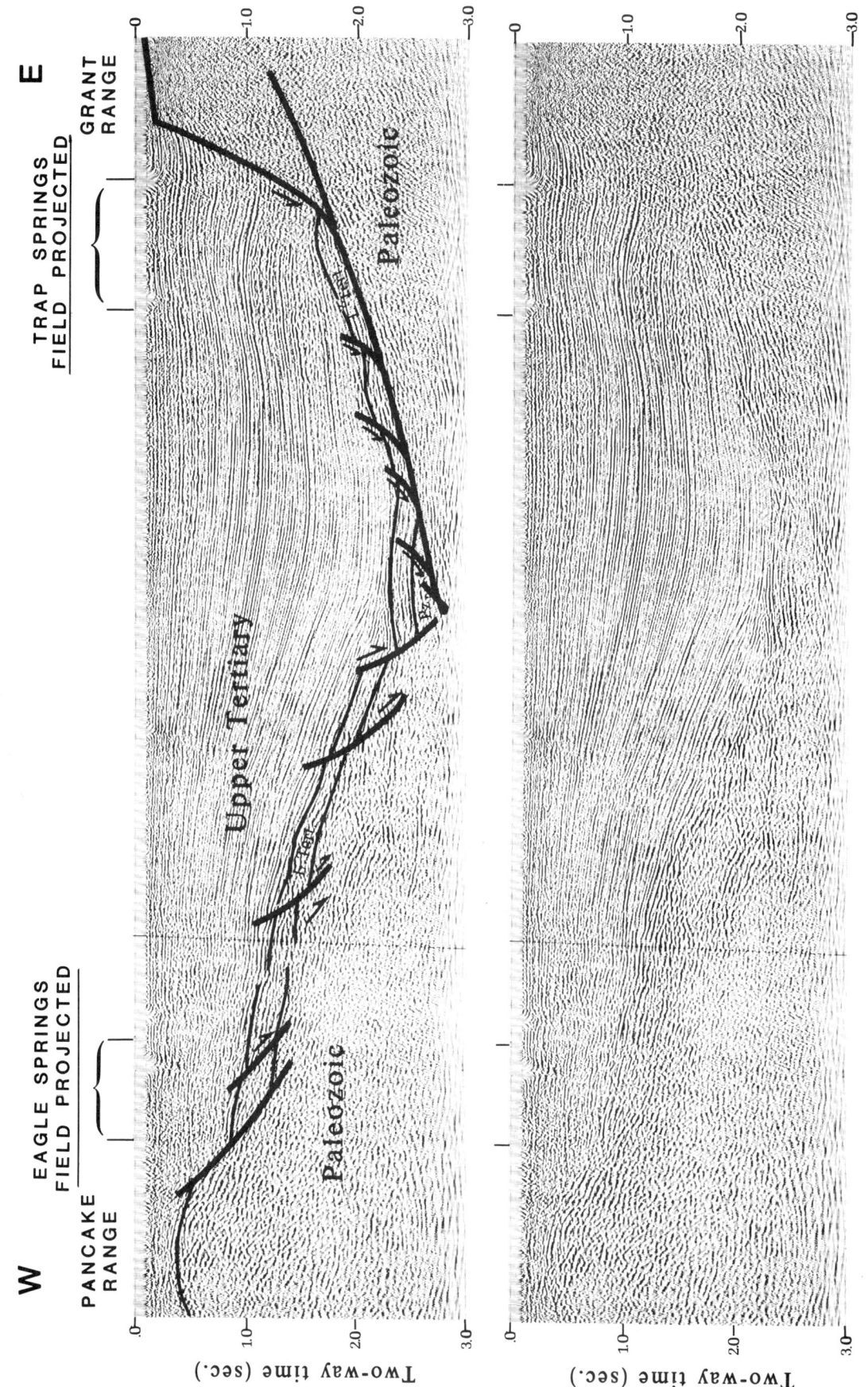

Figure 3. Seismic dip profile for Railroad Valley. Vertical scale is in two-way time. Data is 1200% coverage, stacked, migrated, and with automatic statics applied. Maximum thickness of Tertiary is about 4.5 kilometers.

poor data quality. However, it is expected that the western minor normal faults behave as their eastern counterparts and also pass into the listric fault.

The seismic dip profile in Figure 3 and a number of the elements covered in the description of Railroad Valley can be considered typical of most, but not all, valleys in northeastern Nevada. The four most salient attributes are listed.

1. A westward-dipping listric fault bounds the basin on to the east.

2. A gentle eastward-dipping western flank of the basin is in a hanging wall relation to the listric fault.

3. The asymmetry of the basin is defined by the Tertiary sediments with the basin's depocenter adjacent to the listric fault.

4. The migration of upper Tertiary (Miocene and post-Miocene) depocenters is from west to east through time in response to movement along the listric fault.

DIAMOND VALLEY

Interpretation of the seismic profiles at Diamond Valley (Figs. 5, 6, and 7) is based primarily on the seismic character of reflection packages that can be correlated with stratigraphic units in adjacent basins where both well and seismic data are more abundant. Outcrop data were also used in interpretation of the seismic profiles. Stratigraphic control for the Tertiary section at Diamond Valley is limited to three wells. Seismic data in Diamond Valley contain low-frequency data that make interpretation somewhat less reliable, particularly on smaller sized features.

The seismic dip profiles across northern Diamond Valley and southern Diamond Valley (Figs. 5, 6, respectively) indicate a general similarity to the seismic dip profile across Railroad Valley (Fig. 3). Furthermore, there is a similarity in the overall thickness of the Tertiary sequence that ranges from about 3.0–4.6 km (10,000 to 15,000 ft). In terms of the four salient attributes listed in the previous section, northern Diamond Valley is more similar to Railroad Valley than it is to southern Diamond Valley.

Dissimilarities between the northern and southern Diamond Valley exist primarily in the timing and/or rate of deformation, as well as in their inferred depositional histories. Southern Diamond Valley (Fig. 6) contains a thinner upper Tertiary unit, a thicker lower Tertiary unit than the northern, and it may possibly even contain a Cretaceous unit, which the northern valley lacks. The Cretaceous unit, the Newark Canyon Series, observed in the outcrop of the Diamond Range that flanks the valley to the east may be present in the southern Diamond Valley (Nolan and others, 1956; Stewart and Carlson, 1978). The following sequence of depositional and deformation events can be postulated from the seismic sections. During Cretaceous time the area occupied by southern Diamond Valley was structurally lower than northern Diamond Valley and thus received Newark Canyon sediments. Then the lower Tertiary unit in southern Diamond Valley was deposited but in a more deformed basin than the same unit in northern Diamond Valley. This relationship suggests either an earlier episode of basin formation or a more rapid

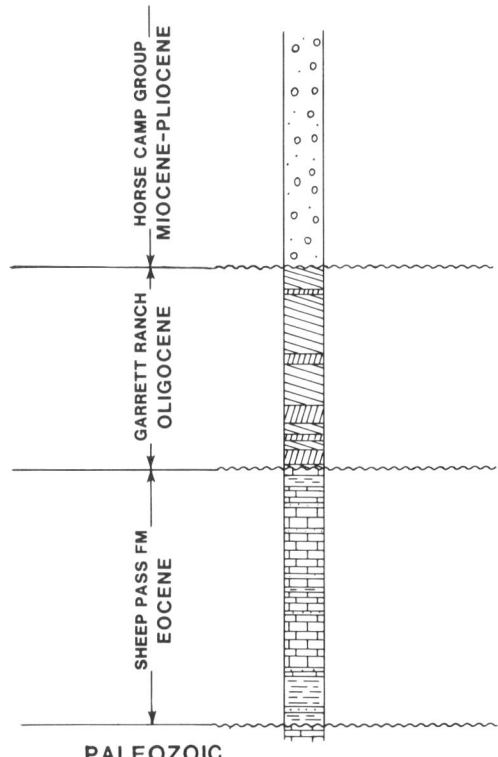

Figure 4. Generalized type section for the Tertiary in Railroad Valley. Horse Camp Group, fluvial and lacustrine clastics; Garrett Ranch Formation, ignimbrites, volcanoclastics and some rhyolitic-dacitic flows; Sheep Pass Formation, lacustrine carbonates and shales with minor clastic components.

history of basin subsidence in the south than in the north during the lower Tertiary. Subsequently, the upper Tertiary units in northern Diamond Valley were deposited and are much thicker than those in the southern Diamond Valley indicating a reversal of subsidence rates between the two valleys.

The differing deformation and stratigraphic histories of the northern and southern portions of Diamond Valley is shown in Figure 7, which is a north-south seismic strike profile. The most prominent feature on the profile is the large mass of Paleozoic rocks located approximately to the left of center on the figure. The Shell Diamond Valley No. 1, drilled in 1956, penetrated this axis and recovered brecciated, chloritized Paleozoic rocks suggestive of a shear zone. In three dimensions, this Paleozoic mass is mapped as an east-west axis separating Diamond Valley into two distinct sub-basins with different timing and/or rate of deformation, as well as differing stratigraphies. The origin of the axis is postulated to be the result of a tear fault associated with the differential movement along the westward-dipping listric fault that, as in Railroad Valley, probably utilized decollement surfaces within the Paleozoic sedimentary section. Differences in the timing and/or rate of movement of the listric fault caused the Paleozoic units in the hanging wall along the tear fault to deform almost chaotically, as suggested by the rocks described from the well that penetrated the axis.

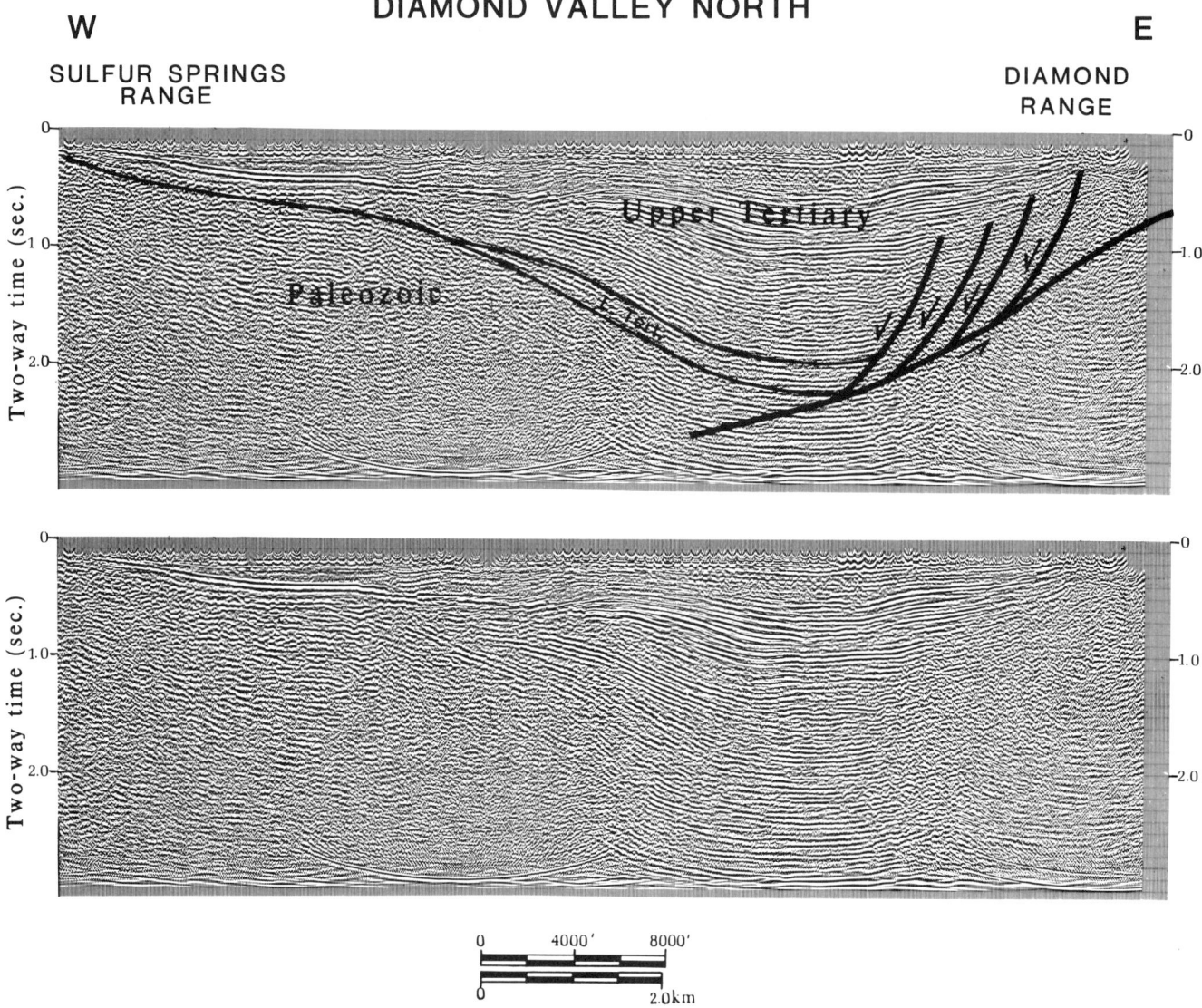

Figure 5. Seismic dip profile for Northern Diamond Valley. Vertical scale is in two-way time. Data is 1200% coverage, stacked, migrated, and with automatic statics applied. Maximum thickness of Tertiary is about 3.15 km (1.96 mi).

Transverse structural axes, as described in Diamond Valley, can be documented for other valleys in northeast Nevada with sufficient geological and geophysical data; however, the subsurface structural relief of the axes vary from valley to valley. Probably in most cases the deformational history of each sub-basin within any given valley will be different. Only more detailed seismic data combined with drilling will provide additional insights to explain such axes.

MARY'S RIVER VALLEY

The seismic dip profile for Mary's River Valley (Fig. 8) exhibits all the four salient attributes of valleys in northeastern Nevada summarized in the section on Railroad Valley. The factor that sets this basin apart from those previously discussed is that the westward-dipping listric fault appears to flatten in a mylonitic crystalline rock at the base of the unmetamorphosed Paleozoic sedimentary sequence, rather than on decollement surfaces within the Paleozoic sedimentary sequence. Reflection fragments are generated from the subhorizontal fault plane on this profile. The description of and possible origin for the mylonite zone are discussed by Misch (1960) and Snelson (1957). The zone located at the base of the Paleozoic section crops out in the Ruby Mountains and adjacent East Humboldt Range that bound Mary's River Valley to the east and can be projected into the seismic profile in the basin (Fig. 8). Stratigraphic control for interpreting the seismic profile in Figure 8 was obtained from well data available updip on the western flank of the basin and from out-

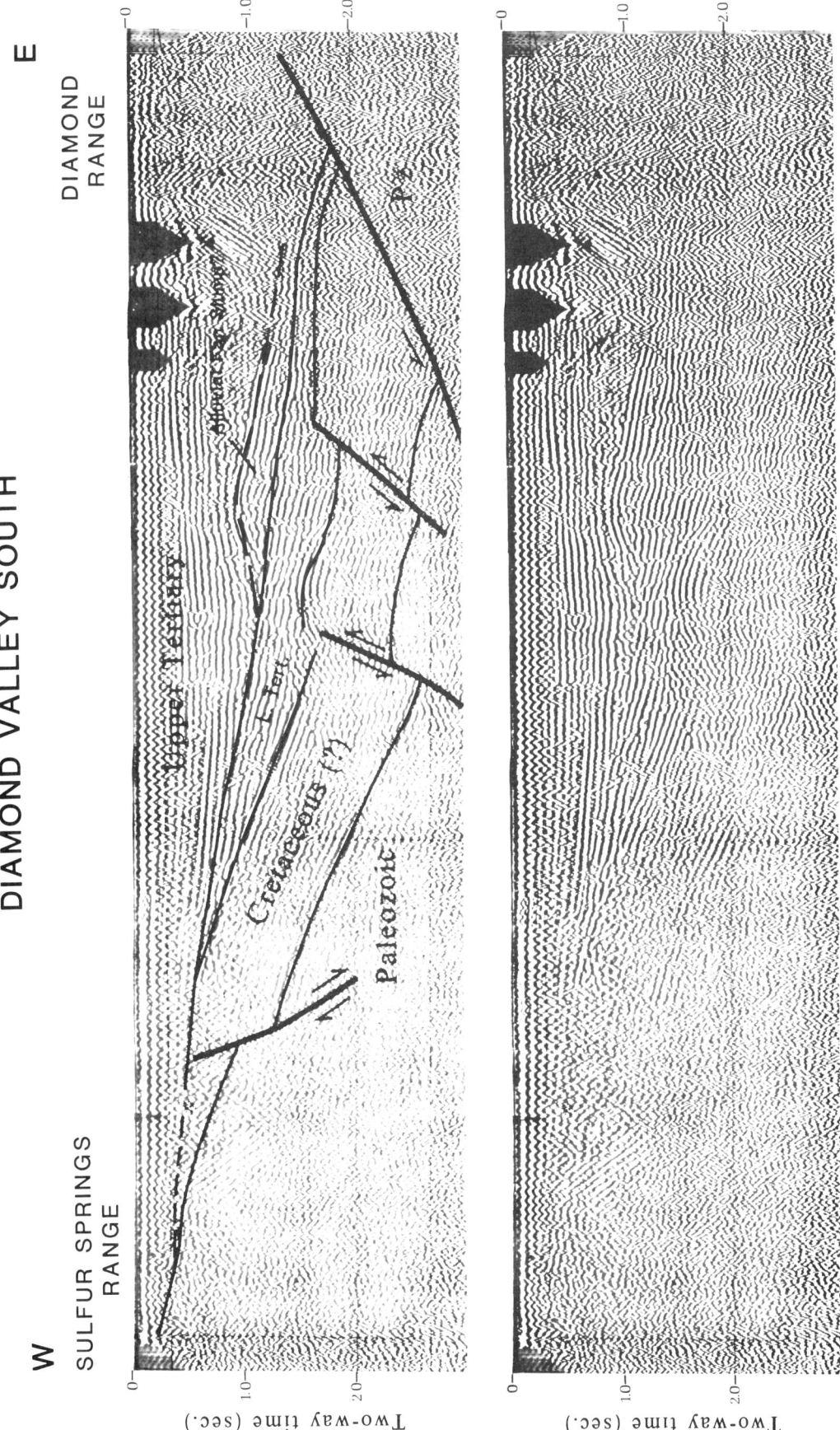

Figure 6. Seismic dip profile for Southern Diamond Valley. Vertical scale is in two-way time. Data is 1200% coverage, stacked, with automatic static applied, and high frequencies recovered. Maximum thickness of Tertiary and Cretaceous (?) is about 4.5 km (2.8 mi).

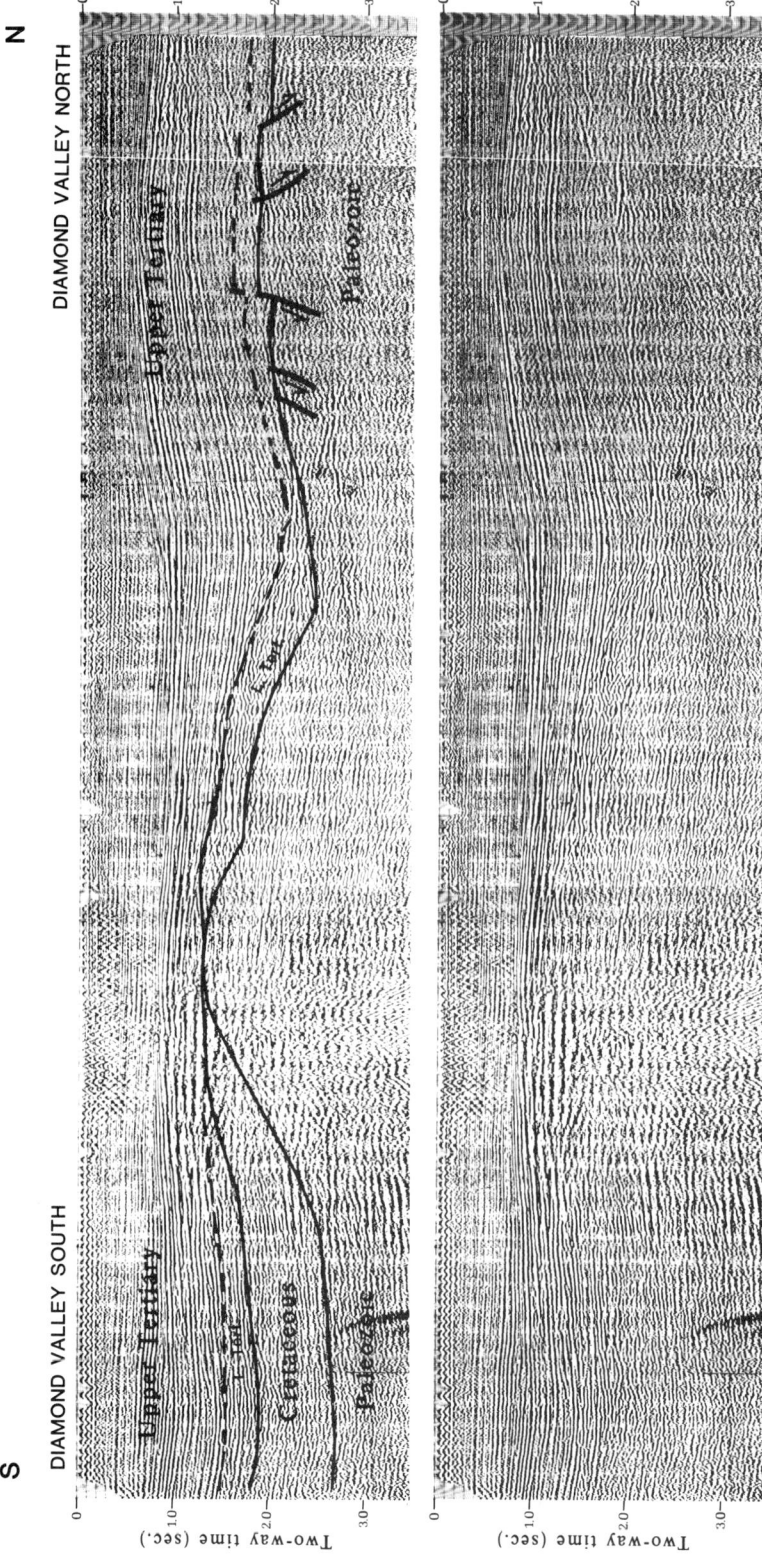

Figure 7. Seismic strike profile for Diamond Valley illustrating transverse axis and differences in stratigraphic units in the adjacent basins. Vertical scale is in two-way time. Data is 1200% coverage, stacked, and with automatic statics applied.

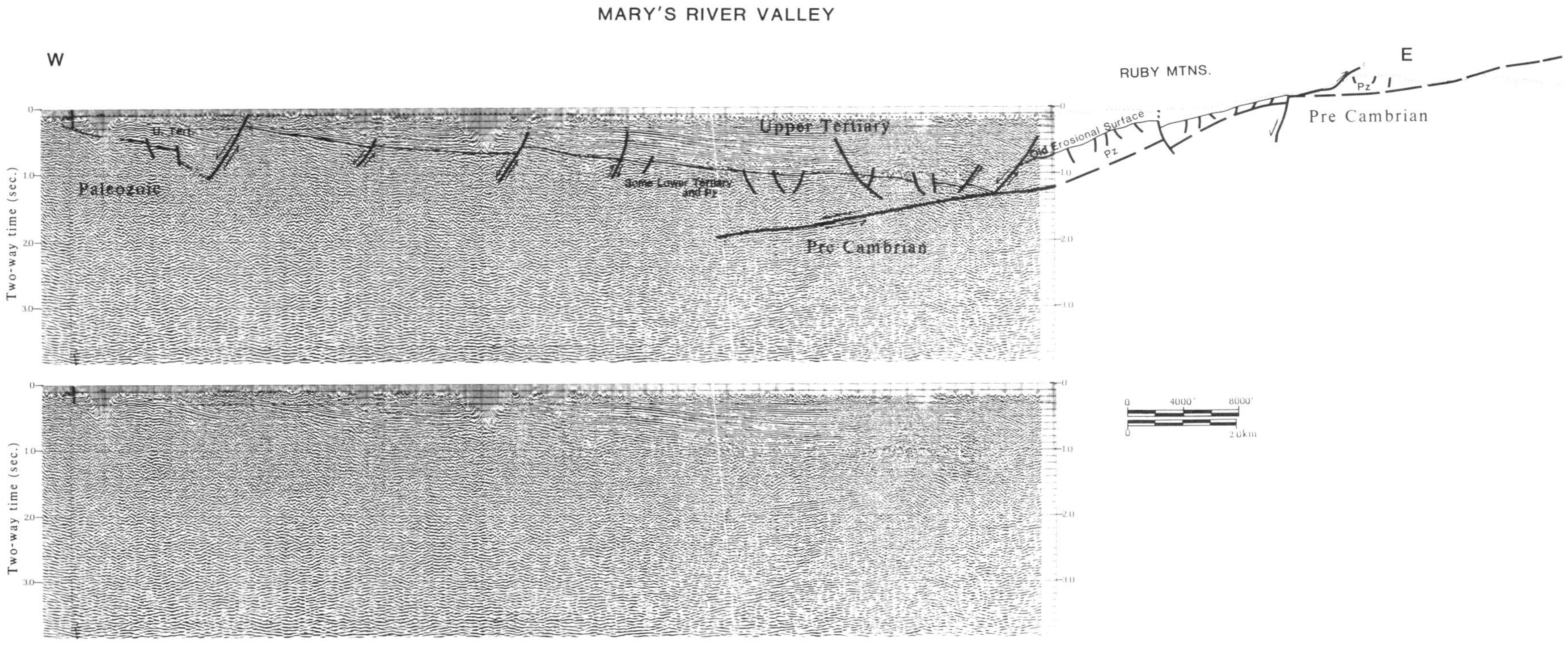

MARY'S RIVER VALLEY

Figure 8. Seismic dip profile for Marys' River Valley. Vertical scale is in two-way time. Data is 1200% coverage, stacked, migrated, and with automatic statics applied. Maximum thickness of Tertiary is about 2.4 km (1.5 mi).

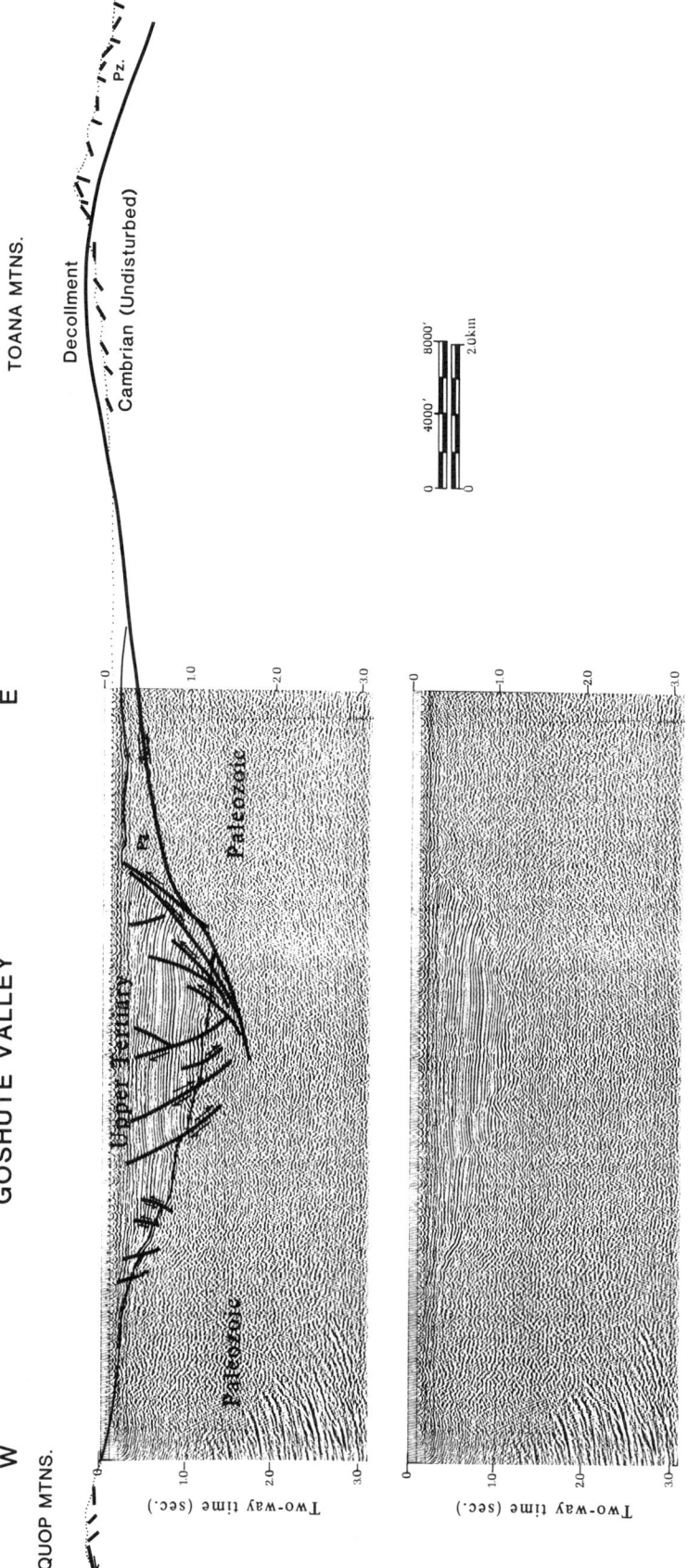

Figure 9. Seismic dip profile for Goshute Valley. Vertical scale is in two-way time. Data is 1200% coverage, stacked, migrated, and with automatic statics applied. Maximum thickness of Tertiary is about 2.7 km (1.7 mi).

crop information (Howard, 1971; Stewart and Carlson, 1978). The absence of good reflecting horizons on the seismic section prevents the separation of the lower Tertiary from the Paleozoic units in the hanging wall, and thus details on the time of basin development initiation are absent. The lack of good reflectors also precludes the seismic determination of the total Tertiary sequence thickness. Magnetotelluric and gravity surveys, however, suggest a maximum depth to the top of the Paleozoic of 2.4 km (8,000 ft).

Minor normal faults observed in Mary's River Valley are of a much younger age (late and post-Miocene) than in the Diamond and Railroad valleys. In these valleys, the upper Tertiary sequence is virtually unfaulted. The minor faults in Mary's River Valley pass into the low-angle listric fault, as also observed in Railroad Valley (Fig. 3).

Robison (1983) confirms the writer's interpretation of the general structural evolution of Mary's River Valley based on additional seismic data and a more detailed stratigraphic analysis.

GOSHUTE VALLEY

Goshute Valley (Fig. 9) is similar to other valleys discussed based on the four attributes previously summarized. However, seismic data indicates that the lower Tertiary present in all the other valleys is absent at Goshute Valley. A well drilled in the center of the basin encountered upper Tertiary units resting directly on Paleozoic rocks. It is therefore suggested that this valley was structurally positive during early Tertiary time, and it may have even served as a source for clastics now found in the adjacent areas. Goshute Valley did not become a basin until late Tertiary time when it accumulated about 2.7 km (9,000 ft) of sediments.

CONCLUSION

Railroad, Diamond, Mary's River, and Goshute valleys in the Basin and Range Province of northeastern Nevada were studied to establish similarities and differences in their Tertiary structural development. Documentation for the study was provided by integrating seismic data with well and outcrop information.

All the valleys contain a Tertiary sequence that, in dip section, defines an asymmetrical basin bounded on the east flank by a major listric fault. Displacement along the listric fault varies from basin to basin but is usually 3 km (10,000 ft), with a range of 2.4 km (8,000 ft) to 4.6 km (15,000 ft). The western flanks of the basins are defined by gently east-dipping ramps. Seismically, the trace of the westward-dipping listric faults are generally interpreted to pass into decollement surfaces in the Paleozoic sedimentary sequence or locally into a mylonite zone at the base of the Paleozoic sedimentary sequence. The decollement surfaces utilized by the listric fault cannot be established in terms of their stratigraphic position for each basin. However, it is possible to speculate that the decollement surface for the listric faulting of Tertiary age is merely a reactivated surface utilized by Mesozoic episodes of thrusting. Such speculation can be justified on the basis of COCORP reflection seismic data collected in the eastern Basin and Range (Allmendinger and others, 1983). The mentioned COCORP data were recorded to produce 20 seconds of stacked data, corresponding to about a 60-km (37-mi) subsurface depth. Seismic penetration to such a subsurface depth is sufficient to indicate potential structural relationships between the shallow listric faulting and the deeper reflections that may be interpreted as fault planes related to Mesozoic thrusting. Seismic data presented in this paper do not have sufficient subsurface penetration to allow for the illustration of the relationship between listric faulting and thrusting; however, seismic data, in conjunction with surface data, may support speculation to that effect.

In each valley, the Tertiary depocenter adjacent to the bounding fault shifted from west to east with continued slippage on the listric fault through time. The greatest rate of movement along the listric fault generally occurred in late Tertiary (Miocene and post-Miocene) time; however, basin formation in most, but not all, valleys was initiated in early Tertiary time. In the strike direction, the valleys are usually separated into two sub-basins by an east-west structurally high axis, postulated to be the result of a tear fault system in the hanging wall associated with differential movement along portions of the listric normal fault.

The thicknesses of Tertiary stratigraphic units vary between basins and between sub-basins in a given valley. All basins contain upper Tertiary rocks; however, not all sub-basins contain the lower Tertiary section, suggesting a complex scheme of structural development in this small portion of the Basin and Range Province.

ACKNOWLEDGMENTS

We would like to express our gratitude to Shell Oil Company for their permission to release for publication the seismic lines and information contained within this paper. Special thanks go to A. W. Bally, who along with S. Snelson, suggested the preparation of this paper and served as constructive critics, and to F. B. Conger who provided many insights in the geology of the region. Particular acknowledgment is due R. Mason who was among the first to make many of the observations reported. We are also grateful to D. Preston, W. Winfrey, and J. Kruger for the many stimulating discussions. Manuscript and illustration preparation was facilitated by E. Lange, T. Matlock, and D. McKinney.

REFERENCES CITED

Allmendinger, R. W., Sharp, J. W., Von Tish, D., Serpa, L., Brown, L., Kaufman, S., Oliver, J., and Smith, R. B., 1983, Cenozoic and Mesozoic structure of the eastern Basin and Range Province, Utah, from COCORP seismic-reflection data: Geology, v. 11, p. 532–541.

Anderson, R. E., Zoback, M. L., Thompson, G. A., 1983, Implications of selected subsurface data on the structural form and evolution of some basins in the northern Basin and Range Province, Nevada and Utah: Geological Society of America Bulletin, v. 94, p. 1055–1072.

Armentrout, J. M., Cole, M. R., and Terbest, H., eds., 1979, Cenozoic paleogeography of the western United States—Pacific Coast paleogeography symposium 3: Los Angeles, Pacific Section, Society of Economic Paleontologists and Mineralogists, p. 335.

Bally, A. W., 1981, Atlantic-type margins, in Bally, A. W., ed., Geology of passive continental margins: History, structure and sedimentologic record: American Association of Petroleum Geologists Education Course Note Series 19, p. 1–48.

Bally, A. W., Gordy, P. L., and Steward, G. A., 1966, Structure, seismic data, and orogenic evolution of Southern Canadian Rocky Mountains: Bulletin of Canadian Petroleum Geology, v. 14, p. 11–40.

Bortz, L. C., and Murray, D. K., 1979, Eagle Springs oil field, Nye County, Nevada, in Newman, G. W., and Goode, H. D., eds., Basin and Range symposium and Great Basin field conference: Rocky Mountain Association of Geologists and Utah Geological Association, p. 441–450.

Duey, H. D., 1979, Trap Spring oil field, Nye County, Nevada, in Newman, G. W., and Goode, H. D., eds., Basin and Range symposium and Great Basin field conference: Rocky Mountain Association of Geologists and Utah Geological Association, p. 469–476.

Eaton, G. P., 1979, Regional geophysics, Cenozoic tectonics, and geological resources of the Basin and Range Province and adjoining regions, in Newman, G. W., and Goode, H. D., eds., Basin and Range symposium and Great Basin field conference: Rocky Mountain Association of Geologists and Utah Geological Association, p. 11–40.

—— , 1982, The Basin and Range Province—origin and tectonic significance: Annual Review of Earth and Planetary Sciences, v. 10, p. 409–440.

Effimoff, I., and Pinezich, A. R., 1981, Tertiary structural development of selected valleys based on seismic data; Basin and Range Province, northeastern Nevada: Philosophical Transactions of the Royal Society of London, ser. A., v. 300, no. 1454, p. 435–442.

Fouch, T. D., 1977, Sheep Pass (Cretaceous? to Eocene) and associated closed basin deposits (Eocene and Oligocene?) in east-central Nevada—implications for petroleum exploration: American Association of Petroleum Geologists Bulletin, v. 61, p. 1378.

—— , 1979, Character and Paleogeographic distribution of upper Cretaceous(?) and Paleogene nonmarine sedimentary rocks in east-central Nevada, in Armentrout, J. M., Cole, M. R., and Terbest, H., eds., Cenozoic paleogeography of the western United States—Pacific Coast paleogeography symposium 3: Los Angeles, Pacific Section, Society of Economic Paleontologists and Mineralogists, p. 97–111.

Howard, K. A., 1971, Paleozoic metasediments in the northern Ruby Mountains, Nevada: Geological Society of America Bulletin, v. 82, p. 259–264.

Kirkpatrick, D. H., 1960, Structure and stratigraphy of the northern portion of the Great Range, east-central Nevada, in Boettcher, J. W., and Sloan, W. W., Jr., eds., Geology of east-central Nevada: Intermountain Association of Petroleum Geologists, 11th Annual Field Conference, Guidebook, p. 186–188.

Lowell, J. D., Genik, G. J., Nelson, T. H., and Tucker, P. M., 1975, Petroleum and plate tectonics of the southern Red Sea, in Fischer, A. G., and Judson, S., eds., Petroleum and global tectonics: Princeton, New Jersey, Princeton University Press, p. 129–153.

McDonald, R. E., 1976, Tertiary tectonics and sedimentary rocks along the transition: Basin and Range Province to Plateau and Thrust Belt Province, Utah, in Hill, G. J., ed., Geology of the Cordilleran Hingeline: Rocky Mountain Association of Geologists, p. 281–317.

Miller, D. M., Todd, V. R., and Howard, K. A., eds., 1983, Tectonic and stratigraphic studies in the eastern Great Basin: Geological Society of America Memoir 157, 327 p.

Misch, P., 1960, Regional structural reconnaissance in central-northeast Nevada and some adjacent areas: observations and interpretation, in Boettcher, J. W., and Sloan, W. W., Jr., eds., Geology of east-central Nevada: Intermountain Association of Petroleum Geologists, 11th Annual Field Conference, Guidebook, p. 17–42.

Montadert, L., Roberts, D. G., de Charpal, O., Guennoc, P., 1979, Riffing and subsidence of the northern continental margin of the Bay of Biscay, in Montadert, L., and Roberts, D. G., eds., Initial reports of the Deep Sea Drilling Project: Washington, D.C., U.S. Government Printing Office, v. 48, p. 1025–1060.

Newman, G. W., 1979, Late Cretaceous(?)–Eocene faulting in east-central Basin and Range, in Newman, G. W., and Goode, H. D., eds., Basin and Range symposium and Great Basin field conference: Rocky Mountain Association of Geologists and Utah Geological Association, p. 167–175.

Newman, G. W., and Goode, H. D., eds., 1979, Basin and Range symposium and Great Basin field conference: Rocky Mountain Association of Geologists and Utah Geological Association, 622 p.

Nolan, T. B., Merriam, C. W., and Williams, J. S., 1956, The stratigraphic section in the vicinity of Eureka, Nevada: U.S. Geological Survey Professional Paper 276, 77 p.

Petroleum Information Corporation, 1984, Resume 1983—a complete annual review of oil and gas activity in the United States: Petroleum Information Corporation, p. 116.

Robison, B. A., 1983, Low-angle normal faulting, Mary's River Valley Nevada, in Bally, A. W., ed., Seismic expression of structural styles, Volume 2, Tectonics of extensional provinces: American Association of Petroleum Geologists, Studies in Geology Series #15, p. 2.2.2-12–2.2.2-16.

Smith, R. B., and Eaton, G. P., eds., 1978, Cenozoic tectonics and regional geophysics of the western Cordillera: Geological Society of America Memoir 152, 388 p.

Snelson, S., 1957, The geology of the north Ruby Mountains and the East Humboldt ranges, Elko County, northeastern Nevada, [Ph.D. thesis]: Seattle, University of Washington, 268 p.

Stewart, J. H., 1978, Basin-range structure in western North America: a review: Geological Society of America Memoir 152, p. 1–31.

Stewart, J. H., and Carlson, J. E., compilers, 1978, Geologic map of Nevada: U.S. Geological Survey, scale 1:500,000.

Winfrey, W. M., 1960, Stratigraphy, correlation and oil potential of the Sheep Pass formation, east-central Nevada, in Boettcher, J. W., and Sloan, W. W., Jr., eds., Geology of east-central Nevada: Intermountain Association of Petroleum Geologists, 11th Annual Field Conference, Guidebook, p. 126–133.

Zoback, M. L., Anderson, R. E., and Thompson, G. A., 1981, Cainozoic evolution of the state of stress and style of tectonism of the Basin and Range Province, in Vine, F. J., and Smith, A. G., Extensional tectonics associated with convergent plate boundaries: Royal Society of London, p. 189–216.

MANUSCRIPT ACCEPTED BY THE SOCIETY MARCH 4, 1986

Geometry of seismically active faults and crustal deformation within the Basin and Range–Colorado Plateau transition in Utah

Walter J. Arabasz
Dale R. Julander*
Department of Geology and Geophysics
University of Utah
Salt Lake City, Utah 84112

ABSTRACT

Detailed earthquake studies throughout the transition zone between the Basin and Range (BR) and Colorado Plateaus (CP) provinces in central and southwestern Utah provide key observations relevant to (1) the subsurface geometry of seismically active faults, (2) the correlation of diffuse seismicity with geologic structure, and (3) the nature of a transitional stress state between the BR and CP provinces. Important new data in the form of three-dimensional earthquake distributions and numerous fault-plane solutions come from six field experiments in which temporary arrays of up to 13 portable seismographs were deployed to supplement a regional seismic network.

Seismic slip predominates on fault segments of moderate (>30°) to steep dip—at least for small to moderate-sized earthquakes (magnitude <5)—based on both fault-plane solutions and hypocentral distributions. Mean and median dips of seismic slip planes for normal- to oblique-slip fault-plane solutions in the study area and vicinity range from 49° to 57°. No convincing evidence has yet been found for seismic slip on either a downward-flattening or a low-angle normal fault in this region though such faults are known to be present.

Low-angle structural discontinuities in the study area appear to play a fundamental role in separating locally intense upper-crustal seismicity above 6–8 km depth from less frequent background earthquakes at greater depth, down to about 15 km. Diffuse epicentral patterns result from block-interior microseismic slip and from superposed patterns of shallow upper-crustal seismicity and subjacent seismicity. If large surface-faulting earthquakes (magnitude 6½ to 7¾) nucleate at about 15 km depth in the study area, as observed elsewhere in the Intermountain region, then rupture pathways remain to be identified between deep nucleation points and existing surface fault scarps. Effective seismic surveillance will require precise resolution of focal depths to discriminate depth-varying seismicity.

Fifty-three fault-plane solutions provide significant detail for mapping changes in upper-crustal stress orientation across the BR-CP transition. Important observations include (1) the alignment of horizontal principal stresses perpendicular and parallel to the BR-CP boundary, (2) average regional orientation of minimum principal stress within the transition in the 102°–282° direction, and (3) an eastward change through the transition from normal faulting to strike-slip faulting to mixed faulting, including compressional reverse faulting. Intermediate principal stress throughout the western and central part of the transition must be close in value to the maximum principal stress, and these two principal stresses interchange in orientation between vertical and a north-northeast–south-southwest horizontal direction.

---

*Present address: Chevron U.S.A., Inc., 6001 Bollinger Canyon Road, San Ramon, California 94583.

INTRODUCTION

Results from on-going earthquake studies of the transition zone between the Basin and Range (BR) and Colorado Plateaus (CP) provinces in central and southwest Utah provide timely information relevant to the geometry of seismically active structures and to patterns of contemporary crustal deformation in this region of intraplate extension. The study area (Fig. 1) encompasses an active southern segment of the northerly-trending Intermountain seismic belt, which follows the boundary between the relatively thin crust and lithosphere of the BR province and the thicker more stable crust and lithosphere of the CP province and the Middle Rocky Mountains (Smith, 1978). Instrumental seismic monitoring of the Utah region dates from 1962, including operation since 1974 of a modern seismic telemetry network (Arabasz and others, 1979, 1980). Within the study area, however, station spacing typically has been too large for resolving fine details either of the three-dimensional spatial distribution of earthquake foci or associated source mechanisms.

The best observational data yet available for investigating crustal seismicity within the BR–CP transition come from detailed field studies using temporary arrays of portable seismographs to supplement the regional seismic network. In this paper we summarize results from a systematic program of such studies carried out in the study area since 1979 by the University of Utah and motivated by increasing geodynamic interest in the BR-CP transition. Our intent is to emphasize new insights related to (1) the subsurface geometry of active faulting, (2) the correlation of diffuse background seismicity with geologic structure, and (3) spatial variations of regional stress orientation across the transition. We rely on other presentations (McKee and Arabasz, 1982; McKee, 1982; Julander, 1983) for complete detailed description of the body of seismological data that we selectively illustrate and discuss here.

At the outset, two points should be emphasized regarding the general relevance of the observations of this paper to extensional tectonics. First, the coincidence of the Intermountain seismic belt with the BR-CP transition in central and southwest Utah (Fig. 1) implicitly links available earthquake information with structures accommodating deformation that is transitional between basin-range extension to the west, where cumulative horizontal extension e (i.e., $\Delta l/l$ where l = length) over late Cenozoic time may exceed 1, and plateau block-faulting to the east, where e over the same time has likely been much less than 1. For the BR province in the Utah region, only its easternmost margin now exhibits significant background seismicity and can be probed by local earthquake seismology. Some of the best local examples of listric normal faults and low-angle detachments demonstrable from seismic-reflection profiling (McDonald, 1976; Anderson and others, 1983; Allmendinger and others, 1983; Smith and Bruhn, 1984) are outside the main seismic belt in areas where there currently is little background seismicity—despite the presence of late Quaternary and Holocene fault scarps (Bucknam and others, 1980). Within the transition region itself, young normal

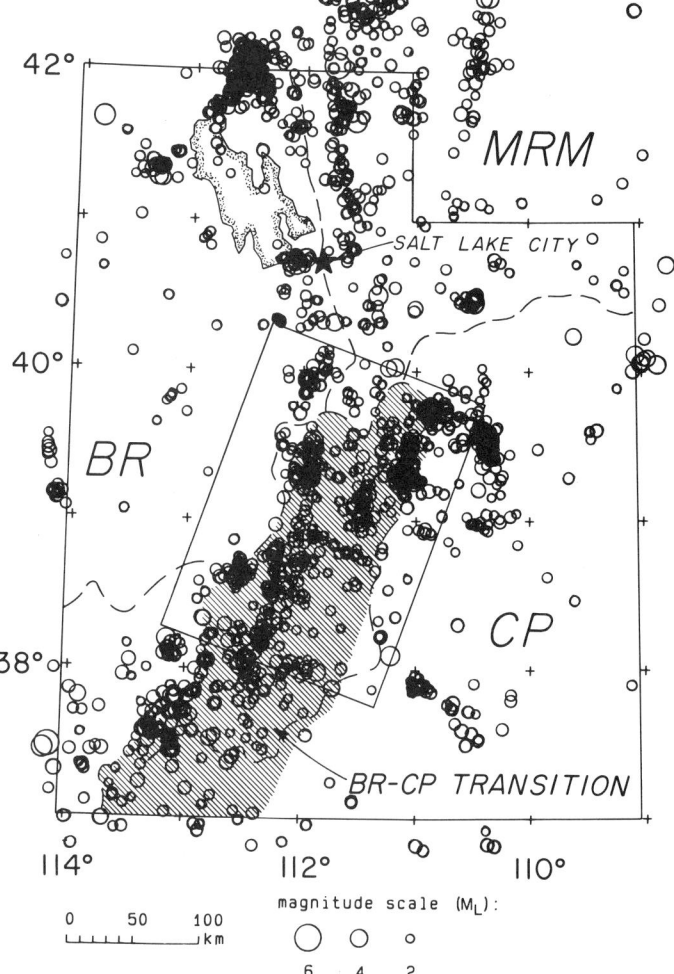

Figure 1. Index map of Utah region showing study area (rectangle) with respect to regional seismicity (1962–1984, M ≥2.0), located by the University of Utah regional seismic network. BR indicates Basin and Range Province; MRM, Middle Rocky Mountains; CP, Colorado Plateaus Province. Alternative outlines of BR-CP transition zone shown by dashed lines (after Stokes, 1977) and by shaded area (after R. E. Anderson, 1985, written communication).

faulting may be more a response to uplift than to lateral stretching (Hamblin, 1984). Moreover, new seismological observations reported here—coupled with results from follow-up geological studies (Anderson and Barnhard, 1984)—emphasize the generally unrecognized importance of young strike-slip faulting within this domain.

Second, most of the instrumental seismicity that we deal with here consists only of small to moderate-sized earthquakes (magnitude <5). (Unless otherwise specified, herein all magnitudes less than 7 are local Richter magnitude; those of 7 or greater are surface-wave magnitude.) There has been no documented instance of historical surface faulting in the BR-CP transition region, and the four largest historical earthquakes (magnitude 6 range) in the transition region occurred before 1922 (Arabasz and others, 1979). Thus, we are handicapped by a lack of information

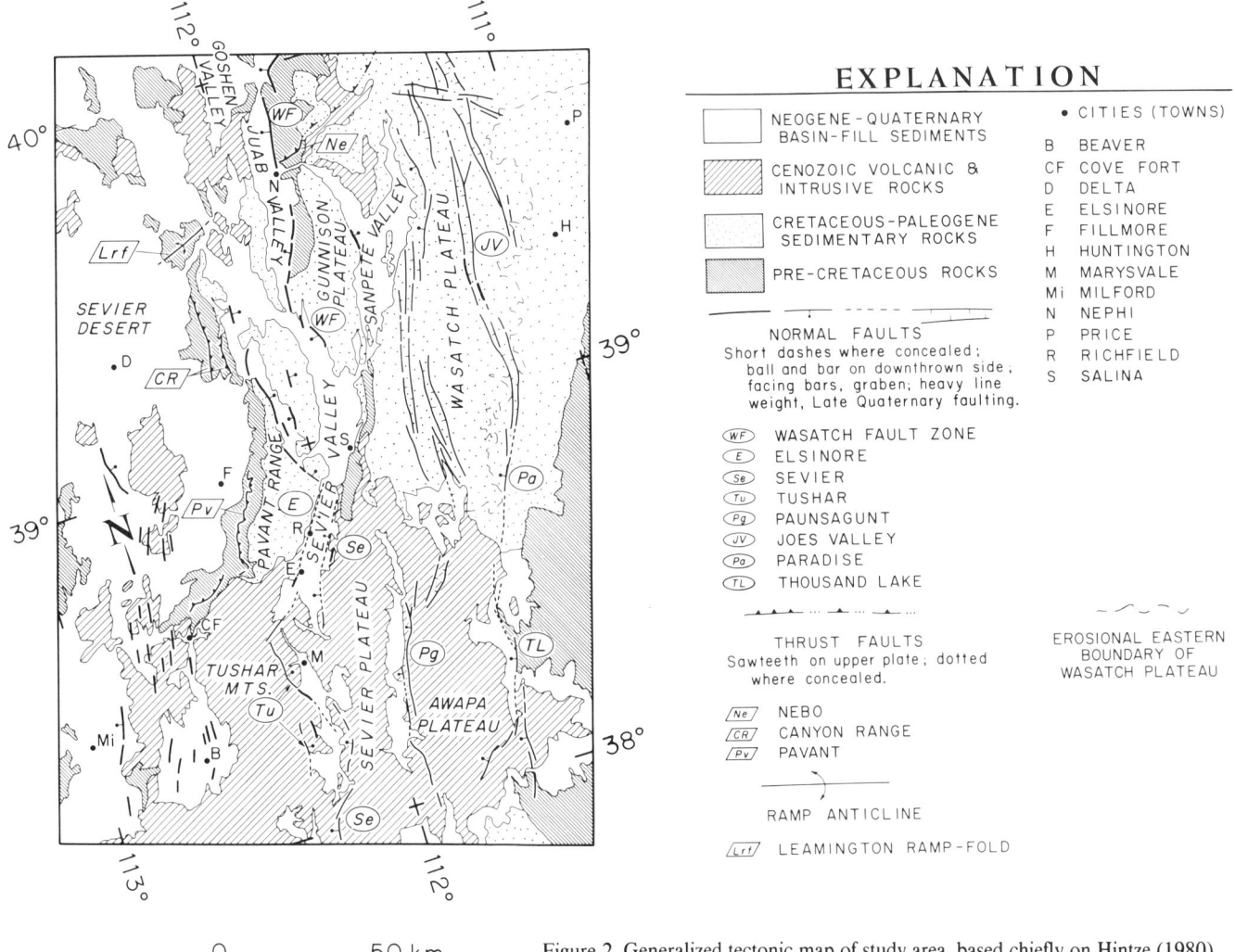

Figure 2. Generalized tectonic map of study area, based chiefly on Hintze (1980).

on large-scale seismic slip related to motion on major faults, and we face uncertainties of dealing with secondary seismic deformation (see Jackson and McKenzie, 1983). In any case, the earthquake observations presented here must be accounted for, and they provide, we believe, important clues to understanding crustal deformation and the generation of episodic large surface-faulting earthquakes in this complex region.

GEOLOGIC FRAMEWORK AND BACKGROUND

The study area (Figs. 1 and 2) encompasses alluviated valleys and fault-bounded mountain blocks of the easternmost BR province and the High Plateaus marking the northwestern rim of the CP province. The boundary between the BR and CP provinces in Utah is well known to be transitional—not only in terms of surface geology and physiography (Spieker, 1949; Burchfiel and Hickcox, 1972; Stokes, 1977; Best and Hamblin, 1978), but also in terms of lithospheric thickness and crustal geophysical parameters (see reviews by Thompson and Zoback, 1979; Keller and others, 1979; Smith, 1978).

The location of the Sevier Valley (Fig. 2) roughly coincides with the beginning of southeastward crustal thickening from BR-type crust (25–30 km thick) to normal CP crust (40 km thick) (Smith, 1978). Thickening of magnetic crust (or, equivalently, the deepening of the Curie isotherm) begins approximately 50 km farther east (Shuey and others, 1973). The associated transition in surface heat flow and its tectonic implications are discussed by Bodell and Chapman (1982; see also Powell and others, 1983). Various criteria may lead to different choices of boundaries for the BR-CP transition as this domain continues to be the subject of active investigation. In Figure 1 we show both the outline of a transition zone formally defined as a physiographic subprovince by Stokes (1977) and boundaries preferred by R. E. Anderson (1985, written communication) on the basis of late Cenozoic geology and faulting.

Within the bounds of the geologic sketch map shown in Figure 2, the classical boundary separating the "Colorado Plateaus Province" from the Great Basin section of the Basin and Range province, as traced by Fenneman (1928, p. 341), follows the western foot of the Tushar and Pavant ranges and then continues northward along the west side of the Gunnison Plateau and the Wasatch Range. The mountain blocks on the west side of the Sevier-Sanpete valley system display transitional character between basin-range and plateau structure. To the east, the Wasatch and Sevier plateaus form part of the true High Plateaus—characterized by high-standing (2,700–3,300 m), north-northeast-trending mountain blocks that are generally flat-topped (capped by Tertiary sedimentary or volcanic rocks) with typically steep sides and intervening valleys ascribed to *en echelon* normal faulting of late Tertiary and Quaternary age (Anderson and Rowley, 1975; Hunt, 1956).

The BR–CP transition coincides with a persistent axis of differential crustal movement since late Precambrian time (Stokes, 1976, 1979). This includes location of (1) the shelf margin of the east-shoaling, Paleozoic Cordilleran miogeocline, (2) the domain where eastward-moving thrust plates broke to the surface producing a reversal of relief during the Cretaceous–early Tertiary Sevier orogeny (Armstrong, 1968), and (3) the eastern limit of late Tertiary–Quaternary basin-range faulting. The eastern margin of Sevier orogenic thrusting has traditionally been linked with the Pavant and Nebo thrusts (Fig. 2). On the basis of regional subsurface exploration, concealed thrust structures associated with Mesozoic-Paleocene thrusting are now traced eastward of the Gunnison Plateau–Sevier Valley area and projected beneath the Wasatch and Sevier plateaus (e.g., Standlee, 1982).

Figure 2 depicts the generalized surface geology, which is dominated in the north by Cretaceous-Paleogene fluvio-lacustrine sedimentary rocks, including synorogenic conglomerates and clastic sediments, and in the south by Tertiary volcanic rocks of the Marysvale volcanic field. The latter, more than 100 km in diameter, is a voluminous complex assemblage of caldera-related flows, breccias, domes, and tuffs resulting from an initial, predominantly Oligocene, phase of intermediate calc-alkaline volcanism and a later, predominantly Miocene, phase of bimodal basalt-rhyolite volcanism (see review by Steven and others, 1984). Although minor in surface exposure, a thick (1.5–2.0 km) sequence of Middle and Upper Jurassic marine-basin strata consisting chiefly of limestones, shales, and evaporites underlies a significant part of the transition region north of the Marysvale volcanic center. Standlee (1982) provides detailed discussion relating these incompetent strata to upper-crustal deformation and structural complexity in central Utah, and argues against widespread salt diapirism (e.g., Witkind, 1982) as an explanation for the same.

Structure in the BR-CP transition should expectedly reflect the complicated superposition of: (1) Mesozoic-Paleocene thrusting and compressional foreland deformation (150–60 m.y. ago); (2) partly overlapping compressional deformation of the Laramide orogeny (75–40 m.y. ago), whose effects extended to monoclinal flexuring within the CP province (Davis, 1978)—and perhaps to two-stage deformation including north-northeastward translation of the Colorado Plateau between about 58 and 44 m.y. ago (Chapin and Cather, 1981; Gries, 1983); (3) pre-basin-range extension (30–10 m.y. ago) that followed cessation of Laramide compression and pre-dated modern basin-range physiography in the region (Zoback and others, 1981); (4) uplift of the Colorado Plateau and structural differentiation of the BR and CP provinces—apparently beginning 30–25 m.y. ago (Rowley and others, 1978, 1979; Bodell and Chapman, 1982) but reflected by significant vertical movement on some plateau block faults much later (Rowley and others, 1981; Best and Hamblin, 1978); and (5) modern basin-range extension (<10 m.y.) in the northern BR province (Zoback and others, 1981) together with rapid uplift of individual plateau blocks such as the Sevier Plateau (7.6–5.4 m.y. ago; Rowley and others, 1981). The influence of pre-Neogene structure on contemporary tectonics will be discussed below.

Cenozoic structure depicted in Figure 2 is dominated by Neogene and younger normal faulting with a predominant north to northeast trend. The southern part of the 370-km-long Wasatch fault zone, the preeminent structure of late Quaternary tectonics in the eastern BR province, lies in the north-central part of the map area. Although without historic surface rupture, the Wasatch fault displays abundant evidence of recurrent late Pleistocene and Holocene surface faulting (Swan and others, 1980; Schwartz and Coppersmith, 1984). The fault has at least 2.6 km of minimum vertical offset in northern Juab Valley and greater structural relief to the north of the study area (Zoback, 1983, Fig. 6). At its southern terminus, the Wasatch fault consists of a series of *en echelon* fault segments along the southwestern flank of the Gunnison Plateau. Late Quaternary deformation in the form of surface faulting appears to be taken up by irregular faulting a few tens of kilometers to the west and southwest of the southern Wasatch fault.

Late Cenozoic basin-fill thickness reflects, in part, both degree of extension (Gibbs, 1983) and magnitude of normal fault displacements in the study area. Basin fill is 2–3 km thick in the Sevier Desert west of the Pavant and Canyon ranges (Zoback, 1983; Allmendinger and others, 1983), 2 km or more thick in the Beaver Basin west of the Tushar Mountains (Machette, 1982), about 1.2 km or less and 0.5 km or less in Juab and Sanpete valleys, respectively (Standlee, 1982; Zoback, 1983), and about 1.3 km or less in the Sevier Valley (Halliday and Cook, 1980; Brown and Cook, 1982).

East of the Sevier-Sanpete valley system, late Cenozoic normal faulting extends 35–50 km eastward into the Wasatch Plateau and at least 50 km eastward into the southern High Plateaus. Total vertical displacements diminish eastward and are estimated as 1,500–1,800 m on the Sevier fault (Hulen and Sandberg, 1981), 750–900 m on the Joes Valley fault zone (Doelling, 1972), and about 700 m on the Thousand Lakes fault (R. E. Anderson, 1984, written communication). Along the easternmost Wasatch Plateau, maximum throws on northerly to northeasterly-trending normal faults diminish to a range of a few hundreds to a few tens of meters or less (Doelling, 1972).

The numerous and complex faults extending the full length of the Wasatch Plateau define a series of *en echelon* north-south-trending graben that have been assumed to be related to post-Eocene subsidence of the Sanpete-Sevier valley system or to localized uplift of the Wasatch Plateau (Spieker, 1949). CO-

CORP seismic-reflection profiling across the western part of the Wasatch Plateau suggests that the plateau faults are probably surficial, not appearing to penetrate the pre-Jurassic section (R. W. Allmendinger, 1984, oral communication). This would be consistent with the inferred presence of a low-angle detachment beneath the Wasatch Plateau at about 4–5 km depth within incompetent Jurassic strata (Standlee, 1982, Fig. 14). Extensional westward backsliding on such a detachment that earlier had an eastward sense of thrusting has been invoked by Royse (1983; see also Standlee, 1982, p. 380) to explain the Wasatch monocline, a major flexure forming the west flank of the Wasatch Plateau.

North of about lat. 39°S, the northeast to north-northeast structural trends of southwest Utah gradually give way to the north to north-northwest trends of northern Utah. Patterns of late Quaternary faulting also change at roughly lat. 39°S. Holocene faulting is widespread to the north, while to the south, numerous late Quaternary fault scarps in southwest Utah appear to be marked by a paucity, if not absence, of Holocene displacement (Bucknam and others, 1980). East-west trends affecting the study area relate chiefly to the regional distribution of Tertiary igneous rocks (Steven and others, 1984; Stewart and others, 1977); Rowley and others (1979) discuss possible structural controls.

REGIONAL SEISMICITY

The study area lies squarely astride a southern segment of the Intermountain seismic belt (ISB) (Fig. 1). The ISB is notably characterized (e.g., Arabasz and Smith, 1981) by: (1) a general predominance of normal faulting, with maximum earthquake magnitudes about 7½ to 7¾; (2) moderate background seismic flux, which is lower by a factor of 4–6 than that along the western North American plate boundary; (3) diffuse seismicity with weak correlation with major active faults and with focal depths almost exclusively shallower than 15–20 km; (4) relatively long (>1000 yr) average recurrence intervals for surface faulting; and (5) a historical paucity of large (M >7.0) surface-faulting earthquakes, despite abundant late Quaternary and Holocene fault scarps. We refer the reader to Smith (1978), Arabasz and Smith (1981), and Smith and Bruhn (1984) regarding aspects of the regional seismotectonics of the ISB, and to Arabasz and others (1980) and Zoback (1983) regarding the seismicity and Cenozoic tectonics, respectively, of the Wasatch Front area adjoining the study area to the north.

Figure 3 shows the map pattern of approximately 2,000 earthquakes within the study area during the period October 1, 1974, to June 30, 1984, based on data from the University of Utah's regional seismic telemetry network. The map includes all earthquakes of magnitude 2.0 or greater (~600) that occurred during the 9.75-year interval. Figure 3 shows the location of epicenters for all earthquakes of magnitude 4.0 since 1962 (see Arabasz and others, 1979; Richins and others, 1981a, 1984), and all earthquakes of estimated magnitude 5.0 or greater since 1850 have been added and specifically indicated. As apparent in Figures 1 and 3, the ISB exceeds 100 km in width in the study area.

At roughly lat. 39°N, following regional structure and physiography, the seismic belt changes orientation from a north-south trend in northern and central Utah to a northeast-southwest trend in southwest Utah.

A first-order feature of Figure 3 is diffusely scattered seismicity throughout the transition, but with intense local clustering that predominantly reflects cumulative background seismicity rather than isolated temporal sequences. Wechsler (1979) made extensive efforts to refine the precision of regional earthquake epicenters through parts of the study area and verified a scattered regional pattern. Most of the epicenters plotted in Figure 3 probably have a precision of ±3 km, based on (1) error analysis (Arabasz and others, 1980), (2) comparison of epicenters located independently by the fixed regional network and by local portable arrays, and (3) reasonably good azimuthal station control (see Richins and others, 1984). Epicentral accuracy of ±5 km would generally be conservative, but errors as large as ±10 km cannot be ruled out at the fringes of the main seismic belt. At the scale of Figure 3 the epicenters should provide a reliable depiction of the map projection of earthquake activity—but an inadequate depiction for confident correlation with subsurface geology. Focal-depth control is generally poor, as usual when the distance to a nearest recording station greatly exceeds the earthquake depth. More than three-fourths of the earthquakes in Figure 3 were located greater than 15 km from the nearest recording station. Referring to Figure 3, from north to south, the following observations can be made with respect to the earthquake map pattern.

Along the Wasatch fault zone, one of the few places where there appears to be any prominent associated seismicity is at its southern end (see Arabasz and others, 1980). However, southwest of station LVU the apparent clustering of earthquakes at small scale close to the fault trace belies simple association with the Wasatch fault, as we later discuss. Clustered epicenters west of station SUU reflect episodic earthquake sequences in Goshen Valley (McKee and Arabasz, 1982). A 20- to 30-km-long segment of the Wasatch fault north of Nephi, possibly associated with the fault's youngest (<300–500 yr?) surface rupture (Schwartz and Coppersmith, 1984), appears relatively aseismic. The broad distribution of small-magnitude earthquakes west of the southernmost Wasatch fault is noteworthy. Curiously, this dispersed seismicity between the Canyon Range and the Gunnison Plateau is roughly bounded on the northwest by the Leamington ramp-fold, an important northeast-striking element of the pre-Neogene thrust-belt structure (Smith and Bruhn, 1984).

A prominent feature of the seismicity of east-central Utah is an intense clustering of epicenters in a horseshoe-shaped pattern open to the south (Fig. 1) that is associated with underground coal mining along the eastern side of the Wasatch Plateau and the arcuate Book Cliffs escarpment. The upper-right corner of Figure 3 includes the western half of the pattern. Microearthquake studies (McKee and Arabasz, 1982; McKee, 1982; Smith and others, 1974) suggest that strain release occurs as shallow induced seismicity a few kilometers beneath the level of mine workings (dis-

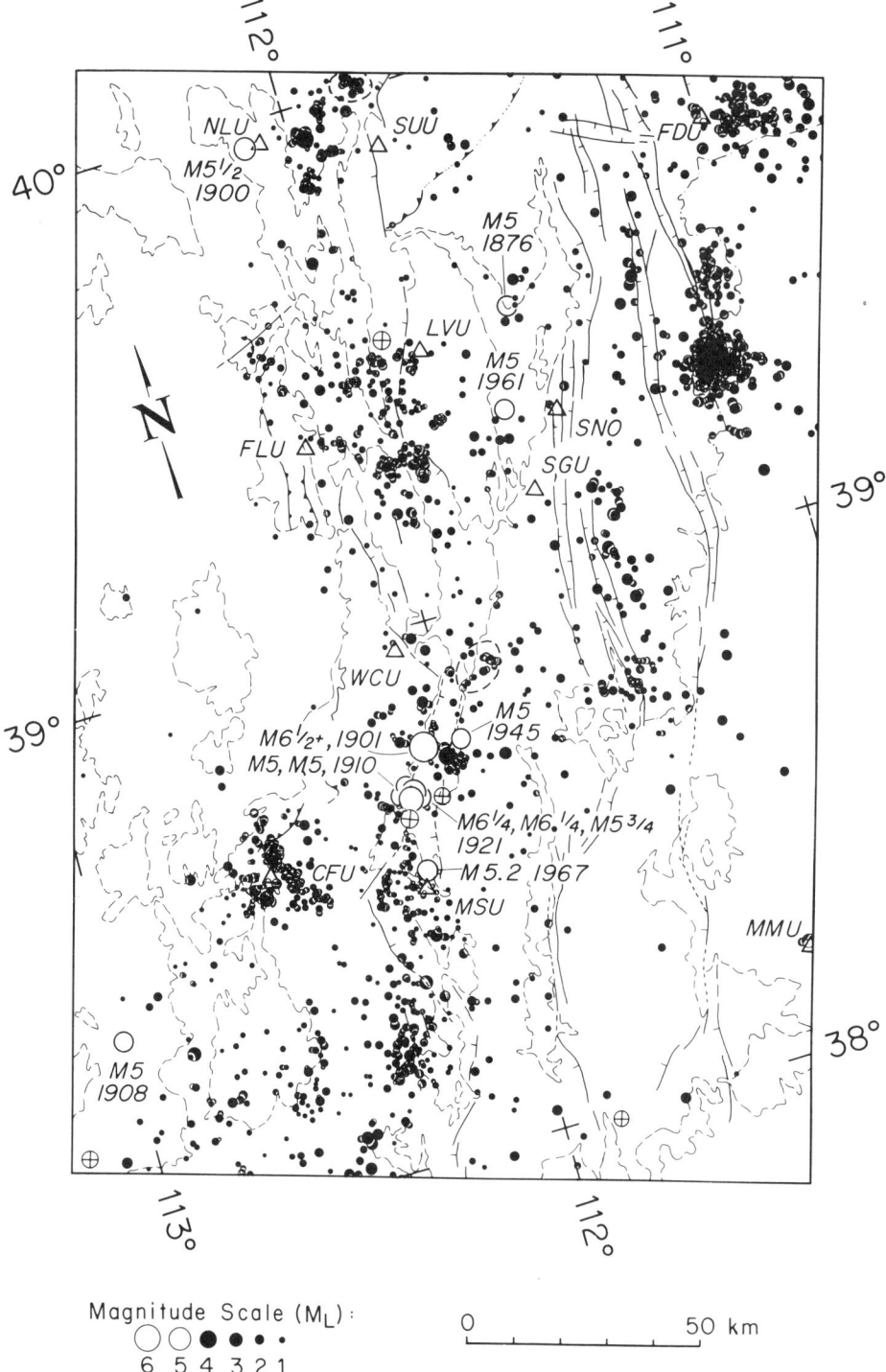

Figure 3. Seismicity map of study area showing instrumental epicenters (solid circles) for the period October 1, 1974–June 30, 1984, based on monitoring by the University of Utah (triangles with three-letter code indicate stations of regional seismic network). Also shown are epicenters for prior earthquakes of magnitude 4.0–4.9 since July 1962 (open circles with inscribed crosses) and epicenters for all shocks of magnitude 5.0 or greater since 1850 (specially labeled open circles). Clustered epicenters for 1910 and 1921 have identical assigned locations at Elsinore, Utah. Dashed ellipses circumscribe two areas where local blasts have probably not been effectively removed from catalog.

cussed later). Seismic events, presumably mining-related, as large as magnitude 4.5 have occurred in the Book Cliffs area (Cook and Smith, 1967).

Throughout the intervening area between the Wasatch fault and the eastern Wasatch Plateau, earthquake epicenters appear randomly scattered. Epicentral density increases, however, in the vicinity of the northern Joes Valley fault zone—as it does in the vicinity of fault zones in the southwestern Wasatch Plateau. It remains to be determined whether this seismicity correlates with shallow faulting of post-Jurassic strata or with deep (>4 km) underlying structure (discussed later).

In the lower half of Figure 3 seismicity is shown to predominate along and west of the Sevier Valley. Earthquake activity extends westward to the eastern margin of the Great Basin (as it generally does throughout the area of Fig. 3) and diminishes eastward beyond the Paunsagunt fault zone. Interior to the Colorado Plateau, small to moderate-sized earthquakes (M ≤3.6) occurred in late 1978 to early 1980 in the Capitol Reef area, a few tens of kilometers east of the lower-right corner of Figure 3 (Humphrey and Wong, 1983). Prior to a magnitude 4.0 earthquake southeast of Richfield in May 1982, microseismicity along the Sevier Valley had predominated beneath its western flanking highlands except for clustering beneath the Antelope Range near station MSU. We return later to the seismicity of the Sevier Valley. Scattered epicenters in the lower-left corner of Figure 3 are part of a broad belt of diffuse seismicity extending into southwest Utah (see Fig. 1). Locally, clustering along the Beaver Basin–Tushar Mountains boundary is apparent.

Earthquake swarm activity, the clustering of earthquakes of similar size in space and time without an outstanding mainshock, is a common feature of southwest Utah (Arabasz and others, 1979, p. 425; Richins and others, 1981b; see also Smith and Sbar, 1974). These earthquake swarms have tended to occur in and near areas of Quaternary volcanism and have had maximum earthquake magnitudes in the upper magnitude 4 range or less. Swarm earthquakes in the study area have occurred episodically in the Cove Fort–Sulphurdale geothermal area (near station CFU, Fig. 3) (Olson, 1976), as reflected by intense epicentral clustering there, and close to the nearby Roosevelt Hot Springs geothermal area (Zandt and others, 1982). Elsewhere in the study area historical earthquake sequences have been characterized by a mainshock or paired mainshocks—although swarm-like sequences of microearthquakes (M<3) have been recorded in places such as Goshen Valley and near the southern Wasatch fault.

Eight of the twelve historical earthquakes of magnitude 5 or greater in the study area have occurred within or very close to the central Sevier Valley (Fig. 3). The largest, a shock of magnitude 6½+ (Modified Mercalli epicentral intensity I_o = IX), struck the Richfield area on November 13, 1901. During September 29–October 1, 1921, following 2½ weeks of foreshock activity, the Elsinore area 10 km southwest of Richfield was struck by two earthquakes of magnitude 6¼, separated by 50 hours, with an intervening shock of magnitude 5¾ (Pack, 1921). Other instances of paired mainshocks very close to Elsinore include two earthquakes of magnitude 5 (I_o = VI) within 38 hours of each other in January 1910 (Williams and Tapper, 1953), and shocks of magnitude 4.4 and 4.0 within 6 months of each other in 1972 (Arabasz and others, 1979). This pattern of closely-timed earthquake doublets is similar to a type of earthquake occurrence described by Lay and Kanamori (1981) as being associated with earthquake source zones that have neighboring strong asperities. The precise source regions of such earthquakes near Elsinore have yet to be identified, but the observed pattern of paired mainshocks is important for assessing future earthquake behavior there.

The pattern of regional seismicity within the study area is well established, but its relation to geologic structure not so. Indeed, the diffuse epicentral pattern in Figure 3 is typical of much of the ISB where the correlation of seismicity with structure has been enigmatic. In the following sections we contribute to unraveling this problematic correlation using accurate earthquake hypocenters and numerous fault-plane solutions from our various field studies. Fortunately, we can take advantage of recent information from other geophysical studies to compare earthquakes and subsurface structure on a scale not previously possible for the study area.

DETAILED EARTHQUAKE OBSERVATIONS

General Remarks

The data presented here were obtained in the study area between 1979 and 1982 from field studies typically involving arrays of 10 to 13 portable seismographs supplementing the University of Utah's regional seismic network. The Goshen Valley study is one exception where fewer seismographs were used. Field and experimental procedures are summarized in the Appendix. As relevant, the variable quality of observations is either discussed in the text or indicated in the figure captions.

A key issue that warrants comment at the outset is that of the correlation of seismicity with geologic structure. Valid correlation clearly requires more than subjective spatial association. Certainly, the accurate resolution of focal depths is a first necessary step to advance beyond subjective association in two-dimensional map view. But even in cross section or three-dimensional view, determining the source mechanism (at least the fault-plane orientation and sense of slip) for one or more earthquakes remains critical for interpreting structural association. For this reason, we place significant emphasis in this paper on attempts to determine reliable fault-plane solutions.

One approach we take related to single-event fault-plane solutions, based on P-wave first motions, for moderate-sized earthquakes recorded by a regional seismic network at distances ranging from several tens to a few hundreds of kilometers. Accurate projection of such data points on a focal sphere depends critically upon focal depth and assumed velocity structure. Where P-wave first arrivals are recorded at near-regional distances exceeding the focal depth (h) by a large factor, the true take-off directions of such waves are significantly more susceptible to location errors on the focal sphere than are those for local earthquakes recorded by an overlying array. In following sections we deal with P-wave first-motion data for two specific magnitude 4 earthquakes recorded since 1980 by the University of Utah's regional seismic network. The focal-sphere plots are based on take-off directions specially provided by J. C. Pechmann (1984, written communication) from a current study of wave propagation and crust-mantle structure in the Utah region.

A second approach deals with the determination of fault-plane solutions from temporary local arrays by inversion of the observed ratios of the amplitudes of P and SV waves (Kisslinger and others, 1981). (SV waves simply correspond to the component of S waves recorded on vertical-component seismographs.) The methodology of SV/P amplitude-ratio inversion is described in the Appendix. Our application of the technique is limited to use of direct, upward-traveling rays, which requires recording at epicentral distances comparable to the focal depth, and to earthquakes of small size such that all amplitudes are recorded on-scale.

We emphasize two points regarding computer algorithms for determining fault-plane solutions based on the amplitude-ratio technique. First, the mathematical inversion is based solely on amplitude measurements and not P-wave first motions. P-wave first motions, however, do provide constraints on acceptable solutions (see Appendix) such that results from the amplitude-ratio inversion are largely but not completely independent of P-wave observations. Second, the final solutions are not necessarily unique. Their value is that each represents a relatively unbiased fault-plane solution that fits observed amplitude measurements in a least-squares sense. Importantly, for our purpose in this paper, the amplitude-ratio technique can provide a single-event fault-plane solution with far fewer stations than would be required to constrain a similar P-wave solution. We can use the amplitude-ratio technique for a type of hypothesis testing by rejecting fault-plane solutions, based on sparse P-wave data, that do not simultaneously satisfy observed amplitude ratios. Also, we can infer significance from the consistency of a number of amplitude-ratio solutions determined in an unbiased way.

Goshen Valley

Goshen Valley provides a useful starting point for our observations because of its setting (1) within the BR province, (2) close to the Wasatch fault, and (3) within the fold-and-thrust-belt domain of allochthonous Paleozoic and Precambrian rocks. Goshen Valley (Fig. 4) is an alluviated, complexly faulted graben composed of a mosaic of several differentially downdropped blocks, gently tilted to the northeast, and with Cenozoic basin fill reaching 1.9 km or more in thickness based on gravity modeling (see Fig. 5 and Davis, 1983). Eastward shifting of the Wasatch fault and transverse topography such as Long Ridge (Fig. 4) reflect structural complexity, inferred to be associated with pre-Cenozoic structure (Zoback, 1983). There is no prominent surficial faulting along the topographic boundaries of Goshen Valley. Buried intrabasin faulting involves steeply-dipping, northerly-trending normal faults and a deeply downdropped central block (see Fig. 5 and Davis, 1983).

Episodic bursts of microseismicity have been detected in Goshen Valley since installation of the University of Utah's telemetered seismic network in 1974. Figure 4, a representative six-year sample of background seismicity, shows broad epicentral scatter and no clear correlation with the surface geology. The scattered seismicity contrasts with quiescence along the Wasatch fault to the south. We earlier noted that this segment ruptured as recently as 300–500 years ago. The northern terminus of surface faulting was close to the present location of station SUU in Figure 4.

Field recording of aftershocks following a magnitude 4.4 earthquake on May 24, 1980, in the western part of Goshen Valley (open circles, Fig. 4) provided good data for investigating at least one of the earthquake sequences in detail. McKee and Arabasz (1982) used data from a fixed set of five seismograph stations shown in Figure 4 as solid triangles to locate 24 aftershocks of magnitude 2.1 or smaller. The data were later processed with a joint-hypocenter-determination (JHD) technique (see Appendix) to improve the *relative* location of events. In Figure 5, we show the resulting locations in cross section, together with subsurface basin structure recently interpreted from gravity by Davis (1983). The mainshock location, computed independently by the regional seismic network, has a focal depth of 7.0 km, a depth routinely assigned where depth control is inadequate in this region. The JHD locations are of good quality. Mean epicentral precision is 1.20 ($\pm$0.49) km, and mean focal-depth precision is 1.27 ($\pm$0.35) km.

In Figure 5 there is an apparent association of the earthquake foci with the planar projection of the down-to-the-east, Goshen Valley fault zone. In map view, McKee and Arabasz (1982) showed that the epicenters for these JHD locations define a north-northwest trend, but that error ellipsoids are also elongate in that same direction, allowing interpretation that the earthquake foci may have only about 3 km of extent along strike. Here, we investigate more critically the focal mechanisms determined by McKee and Arabasz (1982) for these earthquakes in order to probe their structural significance.

Data for the mainshock focal mechanism consist of P-wave first motions recorded by the University of Utah's regional seismic network at distances of 60–330 km. Solution 3 (Fig. 6) is a new fault-plane solution based on revised focal-sphere data provided to us by J. C. Pechmann (1984, written communication). (Throughout this paper, the numbering of fault-plane solutions is consistently keyed to Table 1, which summarizes basic information for each solution.) The mechanism is based on a free-focal-depth solution of 5.3 km, preferred by Pechmann, but virtually identical nodal planes result from a fixed-depth solution of 10.0 km. In other words, Pechmann's methodology does not provide an independent depth discriminant here. Solution 3 indicates (1)

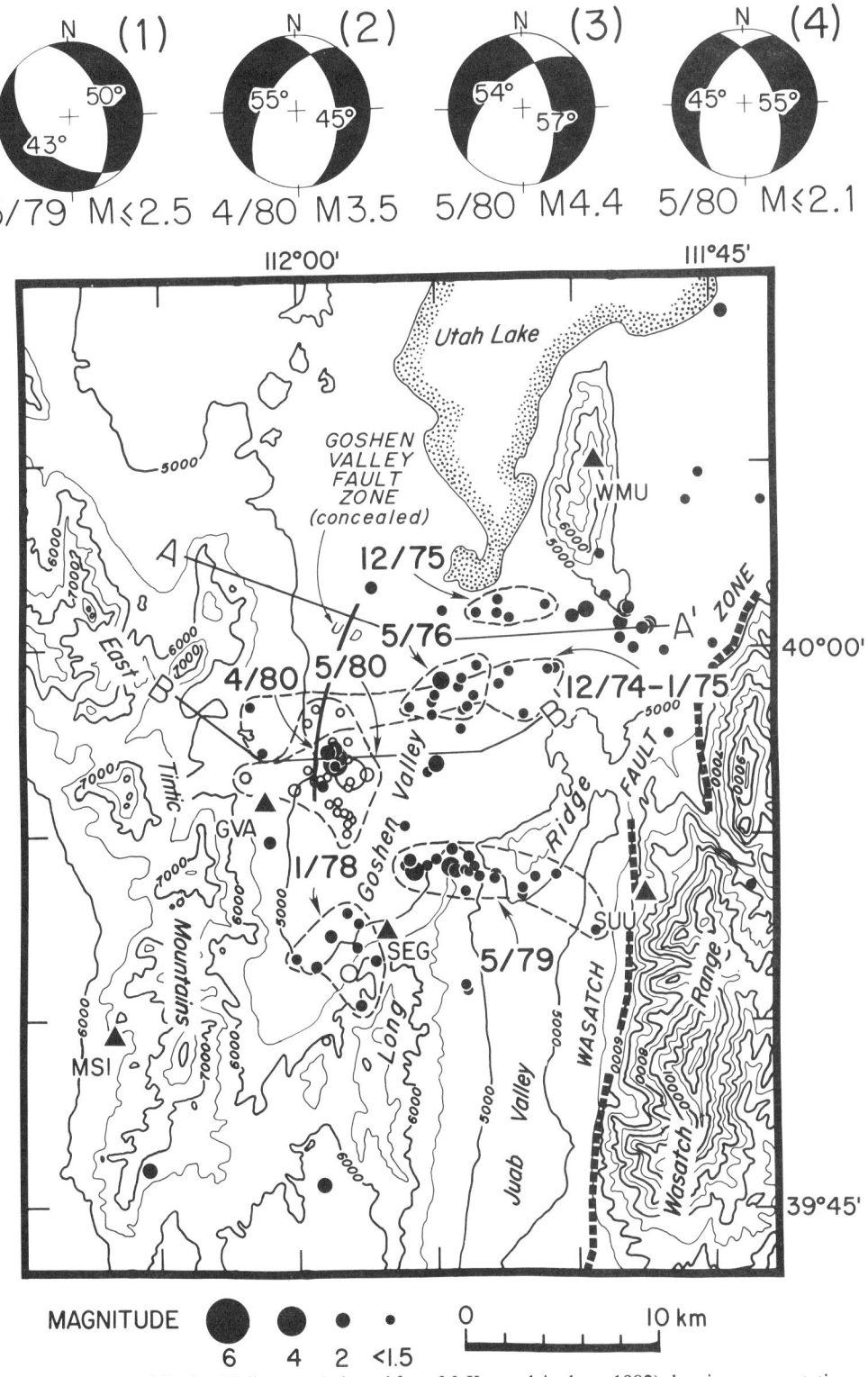

Figure 4. Map of Goshen Valley area (adapted from McKee and Arabasz, 1982) showing representative seismicity (October 1, 1974–September 30, 1980), location of concealed Goshen Valley fault zone (Davis, 1983), and location of cross sections shown in Figure 5. Clusters of epicenters representing temporal bursts of seismicity are circumscribed and labeled by month and year. Corresponding focal mechanisms (keyed to Table 1 and discussed in text) are schematically shown in lower-hemisphere projection, compressional quadrants shaded, at top of figure. Solid circles are epicenters located by the University of Utah regional seismic network; open circles, epicenters located in a special study (see text); triangles, seismograph stations.

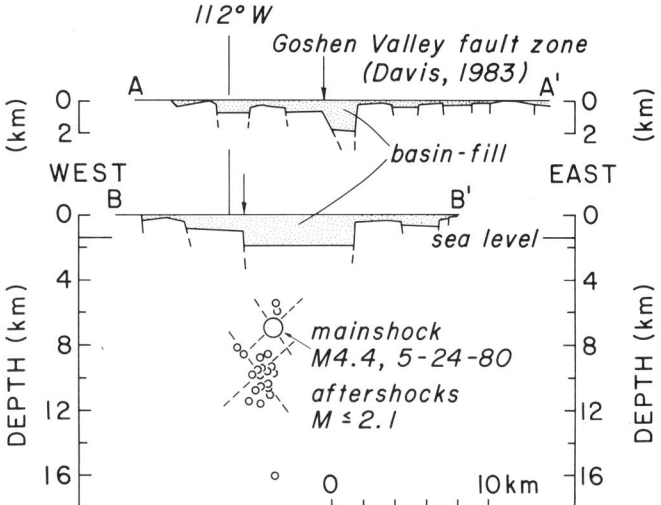

Figure 5. Cross sections (keyed to Fig. 4) showing subsurface basin structure of Goshen Valley interpreted from gravity by Davis (1983), together with earthquake foci (circles) located by McKee and Arabasz (1982). Datum level is 1,470 m above sea level. Dashed lines are the projection of nodal planes from fault-plane solutions 3 and 4 (Fig. 6).

a predominance of normal faulting, (2) a well-constrained nodal plane striking N48°E, dipping 54°NW, and (3) a less constrained nodal plane striking roughly north-south and dipping moderately (44°–57°) to the east. As with many P-wave solutions, the degree of constraint of the northeast-striking nodal plane depends on a few critical data points. Solution 3 differs in detail from that determined by McKee and Arabasz (1982).

Solution 4 (Fig. 6) includes a composite plot of P-wave first motions for a set of the best recorded aftershocks of the May 1980 Goshen Valley earthquake sequence. We have reexamined the same events originally selected by McKee and Arabasz (1982, Fig. 9d). Here, however, we have plotted takeoff directions on the focal sphere based on JHD locations, which were available for 9 of their 10 original earthquakes. Further, we selected one aftershock (h = 8.6 km) with clearly measurable P-wave and SV-wave amplitudes at all stations in order to determine a fault-plane solution using the SV/P amplitude-ratio-inversion technique.

Data points for the single aftershock are represented by the triangles and the X on the stereogram for solution 4 (Fig. 6). The dashed nodal planes are for the corresponding amplitude-ratio solution (solution 4a, Table 1). The solid nodal planes in fact correspond to those originally interpreted by McKee and Arabasz (1982) for the composite of aftershock P-wave first motions. Ninety-three % of the composite P-wave observations are consistent with the solid nodal planes; 70%, with the dashed planes. Accordingly, we prefer the former solution for the aftershocks as a group. Variance between the two, basically similar solutions can readily be explained by slightly variable fracture orientation associated with individual aftershocks.

Similarity in Figure 6 among the single-event amplitude-ratio solution, the composite aftershock P-wave solution, and the mainshock P-wave solution give us confidence that the general source mechanism is correctly established for the earthquake sequence. Schematic stereograms are shown in Figure 4 for solutions 3 and 4 together with two other fault-plane solutions determined by McKee and Arabasz (1982) for earlier earthquakes in Goshen Valley. Although the latter solutions are based on uncritically tested ray directions associated with the regional seismic network, our experience with solution 3 leads us to believe that solutions 1 and 2 would not change significantly.

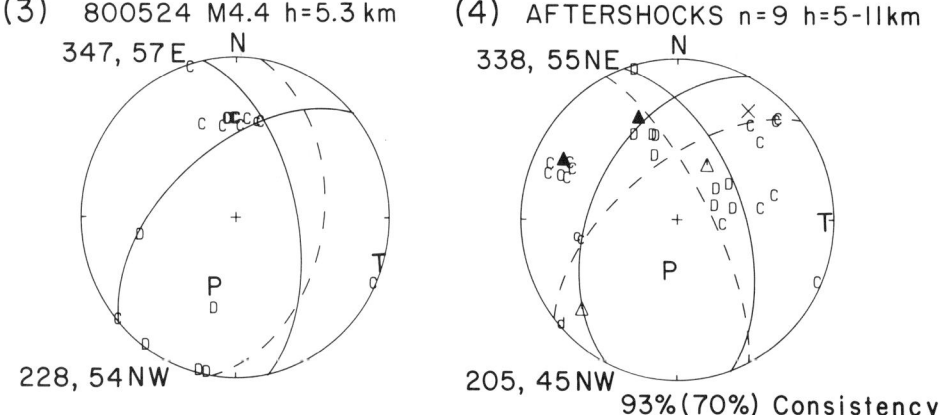

Figure 6. Fault-plane solutions for the May 24, 1980, Goshen Valley earthquake (left) and following aftershocks (right). Projections are lower-hemisphere, equal-area. Triangles represent data points for which both P-wave first motion and SV/P amplitude ratio were used; X's, amplitude data only; C's and solid triangles represent compressions; D's and open triangles, dilatations; P and T, standard pressure and tension axes, respectively. Smaller and lowercase symbols imply lower quality. Dashed nodal planes indicate either range of uncertainty (left) or superposed solution (right) based on SV/P amplitude-ratio inversion (see solution 4a, Table 1). Identification data variously include: solution number, date, magnitude (M), focal depth (h), number of shocks (n) combined in composite, strike and dip of nodal planes (labeled according to strike azimuth for which dip direction is down to the right), and degree of consistency (for composite solutions) between observed data and inferred nodal planes.

TABLE 1. FAULT-PLANE SOLUTIONS

No.[1]	Type[2]	Date YrMoDa	Mag. (M_L)	Nodal Planes Strike[3]	Dip	Rake[3]	P,T-Axes Azimuth	Plunge
1	C(P)	790526	≤2.5	342	50NE	-72	P:314	76
				135	43SW	-111	T:059	05
2	S(P)	800406	3.5	348	45E	-126	P:180	65
				213	55NW	-59	T:282	07
3	S(P)	800524	4.4	228	54NW	-43	P:199	53
				347	57E	-136	T:106	02
4	C(P)	800524-800528	≤2.1	338	55NE	-122	P:190	64
				205	45NW	-52	T:090	06
4a	S(A) rms=0.16	800528	<1.5	231	61NW	-19	P:192	31
				330	73NE	-156	T:097	07
5	C(A) rms=0.25	790624-790726	≤1.7	126	45SW	+162	P:350	21
				229	77NW	+46	T:098	41
6	S(A) rms=0.15	790627	1.0	046	50SE	-78	P:013	78
				208	41NW	-103	T:128	05
7	S(A) rms=0.19	790627	1.5	039	27SE	-13	P:024	46
				140	84SW	-116	T:251	34
8	S(A) rms=0.26	790720	1.6	161	30SW	-156	P:354	50
				049	78SE	-62	T:118	27
9	S(A) rms=0.16	820725	<1.5	268	66NW	-20	P:228	30
				006	72E	.156	T:136	04
10	S(A) rms=0.18	820727	<1.5	013	64E	-127	P:235	56
				253	44NW	-40	T:129	12
11	S(A) rms=0.22	820727	<1.5	162	88SW	+170	P:208	06
				252	79SW	+2	T:116	09
12	C(P)	820725-820727	<1.5	002	75E	-116	P:240	54
				245	30NW	-32	T:112	25
13[4]	S(P)	820524	4.0 (1) 048	65SE	-60	P:359	59
				173	37SW	-137	T:117	16
			(2) 010	80E	-130	P:242	42
				268	40N	-18	T:128	24
			(3) 026	85SE	-90	P:298	50
				206	05NW	-90	T:115	40
14	S(P)	630707	4.4	350	74NE	-127	P:220	48
				238	40NW	-26	T:106	20
15	S(P)	671004	5.2	275	15N	+14	P:247	39
				170	86W	+105	T:094	47
16	S(P)	711111	3.7	345	65NE	-76	P:282	67
				136	30S	-116	T:065	19
17	S(P)	660816	5.6	194	80NW	+171	P:239	00
				285	80NE	+8	T:150	14
18	C(P)	750614-750706	<2.0	338	64NE	-140	P:197	46
				228	54NW	-31	T:110	14
19	C(P)	750630-750702	<2.0	340	14NE	-133	P:126	54
				203	80NW	-80	T:285	34
20	S(A)	810709	<1.0	039	74SE	-21	P:356	27
				135	69SW	-163	T:088	04
21	C(P)	810723-810727	<2.0	352	20E	-62	P:037	62
				143	72SW	-99	T:240	27
22	C(P)	810201-810515	≤4.6	046	50SE	-50	P:024	61
				174	54W	-128	T:289	02
23	C(P)	801221-810121	≤3.4	356	60E	+180	P:217	21
				086	90	+29	T:315	21
24[5]	S(P)	810514	3.5	242	50NW	+65	P:349	03
				100	46S	+118	T:086	70
25	C(A)	790808-790814	≤2.3	319	49NE	+107	P:037	03
				115	43SW	+72	T:299	78
26	C(P)	700723-700826	<3.0	015	20E	+90	P:285	25
				195	70W	+90	T:105	65
27	C(P)	790910	≤2.4	062	60SE	-173	P:281	26
				328	85NE	-30	T:018	18
28	S(P)	791023	3.2	140	50SW	-76	P:106	79
				299	42NE	-106	T:220	05

[1]Source: 1-2, McKee and Arabasz (1982); 3-13, this study; 14-17, Smith and Sbar (1974); 18-19, Olson (1976); 20-21, Zandt and others (1982); 22-23, Richins and others (1981b); 24, Wong (1984); 25, McKee (1982); 26, Smith and others (1974); 27, Wong and Simon (1981); 28, Humphrey and Wong (1983).

[2]S indicates single-event solution; C, composite; (P), based on P-wave first motions; (A), based on SV/P amplitude-ratio inversion. For latter, indicated rms error is for log (SV/P) (see Appendix).

[3]Following Aki and Richards (1980), strike measured here as compass azimuth with fault dipping down to right; rake, as angle (0° to 180°) between strike and slip directions in plane of fault--measured positive counterclockwise for reverse dip slip, and negative clockwise for normal dip slip.

[4]Non-unique solution. Parameters for three alternative models shown in Figure 13.

[5]Schematic focal mechanism only published; parameters are approximate.

Returning to Figure 5, the nodal planes for fault-plane solutions 3 and 4, projected on the section, argue against simple association of the earthquakes with a high-angle planar projection of the Goshen Valley fault zone. Nor can the earthquakes readily be associated with low-angle slip on some listric westward projection of the Wasatch fault. Planar projection of the moderately west-dipping nodal planes in Figure 5, say through point B′, would intersect the surface 4 to 8 km west of the *en echelon* surface traces of the Wasatch fault.

The data in hand provide only ambiguous associations between the Goshen Valley earthquakes and the surface geology. We find no evidence of seismic slip on east-west–trending transverse structures inferred in this locality (Zoback, 1983) and caution that east-west epicentral alignments alone are insufficient proof such structures are seismogenic. Contrasted with what we shall report for other localities in central Utah, the relatively deep focus (6–12 km) of the Goshen Valley seismicity is noteworthy. During an earlier microearthquake study in 1975, Kastrinsky (1977) accurately located clustered microearthquakes 8–12 km deep in the same area as the May 1980 sequence but had inadequate fault-plane control. We obviously cannot conclude that all background seismicity beneath Goshen Valley is similarly deep. We *can* conclude, however, that significant seismicity is occurring either near the base of a plate of allochthonous, highly-faulted Paleozoic and Precambrian rocks—or possibly within an underlying plate (see Hintze, 1980, cross section E–F; Smith and Bruhn, 1984). It remains to be determined whether there is any association between the observed extensional faulting at depth on moderately-dipping (40°–60°) fault planes and the overlying intrabasin block faulting of Goshen Valley (Fig. 5).

Southern Wasatch Fault

Moving into the transition zone itself, we next focus on the vicinity of the southernmost Wasatch fault. Our aim is to combine accurate earthquake locations (Fig. 7) from special studies by McKee and Arabasz (1982) and Arabasz (1984) together with recently published subsurface information (Fig. 8) from Standlee (1982) and Smith and Bruhn (1984). Again, we have made major efforts to determine reliable fault-plane solutions to investigate structural association of the seismicity.

The area of Figure 7 lies within a domain chiefly characterized by Mesozoic-Tertiary sedimentary rocks (Fig. 2). (The pre-Cretaceous rocks indicated along the western flank of the Gunnison Plateau in Figure 2 are Jurassic marine-basin strata.) The character of the Wasatch fault changes significantly as it enters this domain south of Nephi (Schwartz and Coppersmith, 1984; Smith and Bruhn, 1984) such that observations here cannot simply be generalized to other parts of the Wasatch fault. The so-called "Levan segment" shown in Figure 7 is marked by discontinuous late Pleistocene and Holocene fault scarps and has experienced only one surface-faulting event (<1750 years ago) during the past 7300 years (Schwartz and Coppersmith, 1984).

Figure 7 illustrates a pattern of scattered seismicity in the

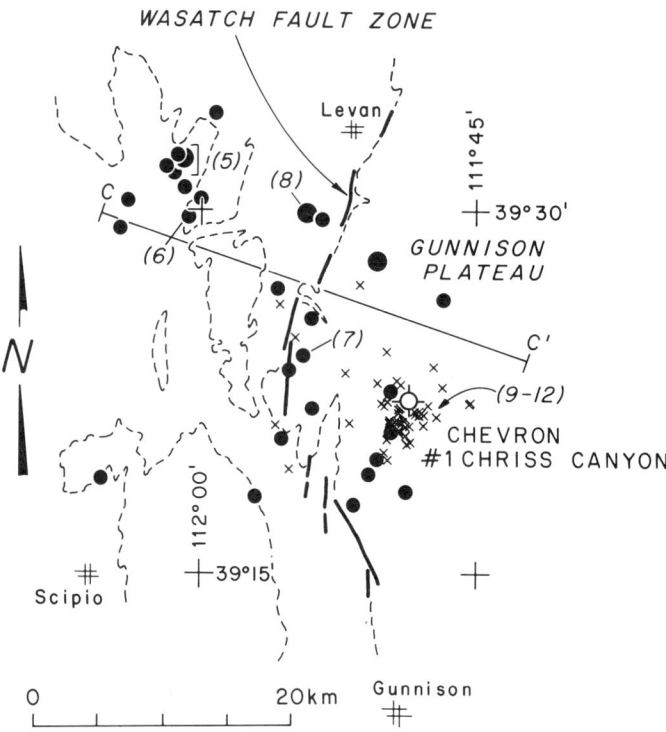

Figure 7. Sketch map of southern Wasatch fault zone showing epicenters of earthquakes located in special studies by McKee and Arabasz (1982); solid circles) and by Arabasz (1984; X's). Magnitudes for smaller circles are less than 1.5; for larger circles, 1.5–1.8; for X's, less than or equal to 2.1. Numbers in parentheses identify earthquakes associated with fault-plane solutions (Fig. 9) and specially indicated in cross section C–C′ (Fig. 8). Mean epicentral precision is 1.23 (±0.72) km for the circles and 1.00 (±0.84) km for the X's.

vicinity of the southernmost Wasatch fault similar to that observed from the regional seismic monitoring (Fig. 3). Epicentral clustering southwest of Levan has been persistent since at least the 1960's. Following a magnitude 4.4 earthquake in Juab Valley in July 1963, Westphal and Lange (1966) carried out brief microearthquake recording. They located scattered aftershocks throughout the southern end of Juab Valley and found a tendency for epicentral clustering along the southwestern side of the valley. Intense epicentral clustering in the vicinity of the Chevron #1 Chriss Canyon well (Fig. 7) relates to swarm seismicity ($M_L \leq 2.1$) fortuitously recorded during a field experiment in June–July 1982 and inferred to have been associated with pore-pressure changes following deep hydraulic fracturing at a depth of 5,070 m in the Chriss Canyon well (Arabasz, 1984). Here we focus on focal mechanisms associated with the swarm seismicity for comparison with those for neighboring seismicity.

The line of section C–C′ (Fig. 7) is collinear with that of a geologic section compiled by Standlee (1982) on the basis of data from drill-holes, seismic reflection profiling, gravity surveys, and surface geology. Figure 8 shows the western half of Standlee's (1982) cross section together with earthquake foci from Figure 7 and key seismic-reflection information from a profile that is

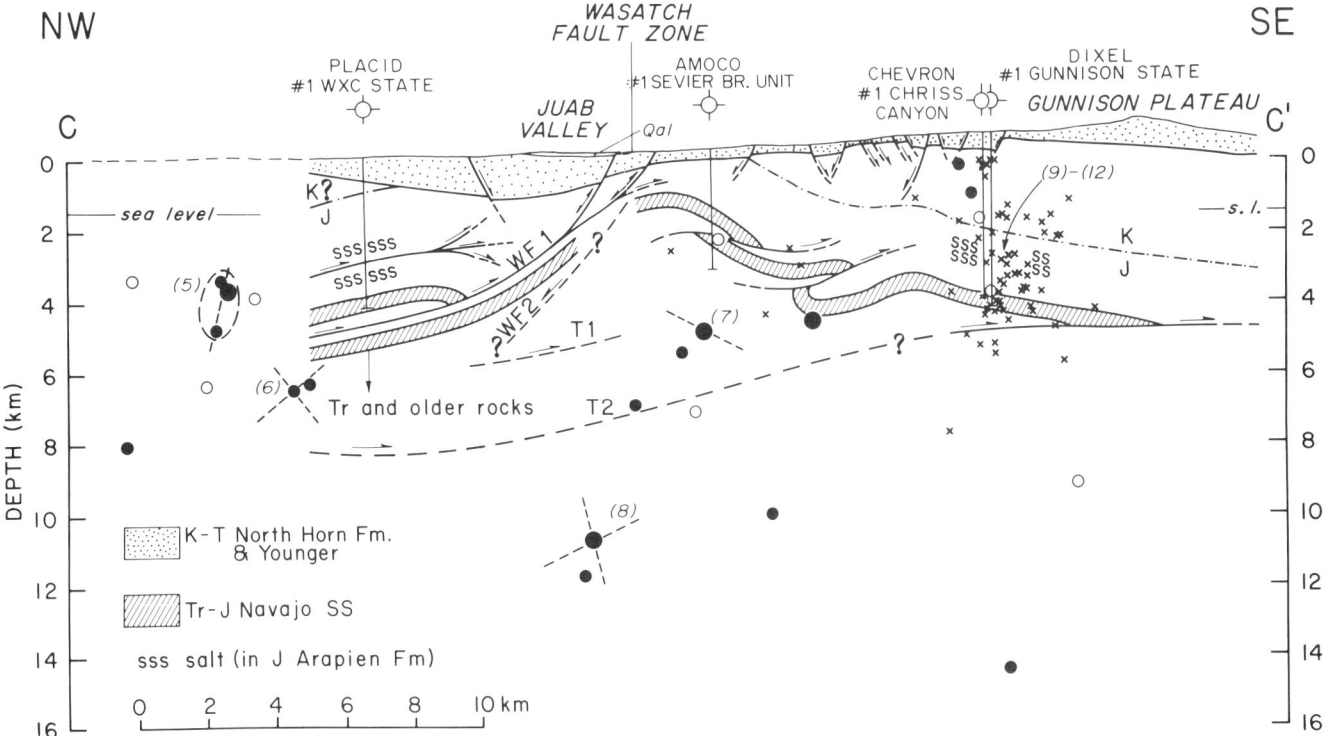

Figure 8. Cross section (keyed to Fig. 7) showing earthquake foci superposed upon interpretations of subsurface geology from Standlee (1982) and seismic-reflection interpretations from Smith and Bruhn (1984). The projection includes earthquakes in Figure 7 within 14 km normal distance. Foci having depth locations with a standard error exceeding 2.5 km are shown as open circles. Datum level is 1,470 m above sea level. WF1 and WF2 are respective interpretations of the Wasatch fault at depth by Standlee (1982) and Smith and Bruhn (1984); T1 and T2, major thrust faults interpreted by Smith and Bruhn (1984). Nodal planes from fault-plane solutions 5–8 (Fig. 9) that strike more than 45° from the line of section have been superposed as dashed lines. Mean focal-depth precision for the solid circles and X's, respectively, is 1.47 (±0.62) km and 1.39 (±1.01) km.

nearly coincident with the line of section interpreted by Smith and Bruhn (1984).

The subsurface shown in Figure 8 is dominated by structures resulting from eastward-directed thrusting during Late Cretaceous to Paleocene time. The Triassic-Jurassic Navajo Sandstone provides a marker bed showing dislocations and thrust-related folding. Immediately east of the Gunnison Plateau, "backthrusting" on east-dipping thrust faults was of major importance in the pre-Neogene, compressional foreland deformation of the region (Standlee, 1982). We are primarily interested in the overprint of Neogene and younger extensional faulting. Both Standlee (1982) and Smith and Bruhn (1984) interpret the data as indicating young normal movement on the preexisting thrust planes. Standlee (1982) depicts the Wasatch fault in this area as a listric normal fault (WF1, Fig. 8) following old thrust surfaces or steeply dipping pre-Tertiary bedding planes. Smith and Bruhn (1984) depict the fault projecting with moderate dip to a slightly greater depth (WF2, Fig. 8) and describe it as either flattening into or terminating at a prominent detachment (T1, Fig. 8).

Earthquake foci shown in cross-section view (Fig. 8) appear randomly scattered except for the swarm seismicity (X's) spatially associated with the Chevron #1 Chriss Canyon well. Excluding the latter, three-fourths of the foci shown as circles lie above the major low-angle detachment labeled T2. Foci at or below 10 km depth should be close to or within Precambrian rocks (McKee and Arabasz, 1982, Fig. 7). An obvious and tantalizing question is whether the earthquakes reflect slip on the numerous low-angle fault planes identified in the cross section. We address the question by examining their source mechanisms.

Figure 9 shows eight fault-plane solutions that we have determined from the seismicity sample. The corresponding earthquakes are labeled both in map and cross-section view in Figures 7 and 8, and resulting fault planes that strike more than 45° from the line of section have been projected onto Figure 8 for the non-swarm earthquakes. Solutions 5–8 (Fig. 9) were determined by applying the amplitude-ratio technique to selected earthquakes located by McKee and Arabasz (1982) and for which P-wave first motions alone were inadequate to constrain nodal planes. For solution 5 we grouped three close earthquakes to obtain adequate focal-sphere coverage; solutions 6–8 are single-event solutions. Solutions 9–12 were determined from the 1982 swarm seismicity: solution 12 as a P-wave composite of the most pre-

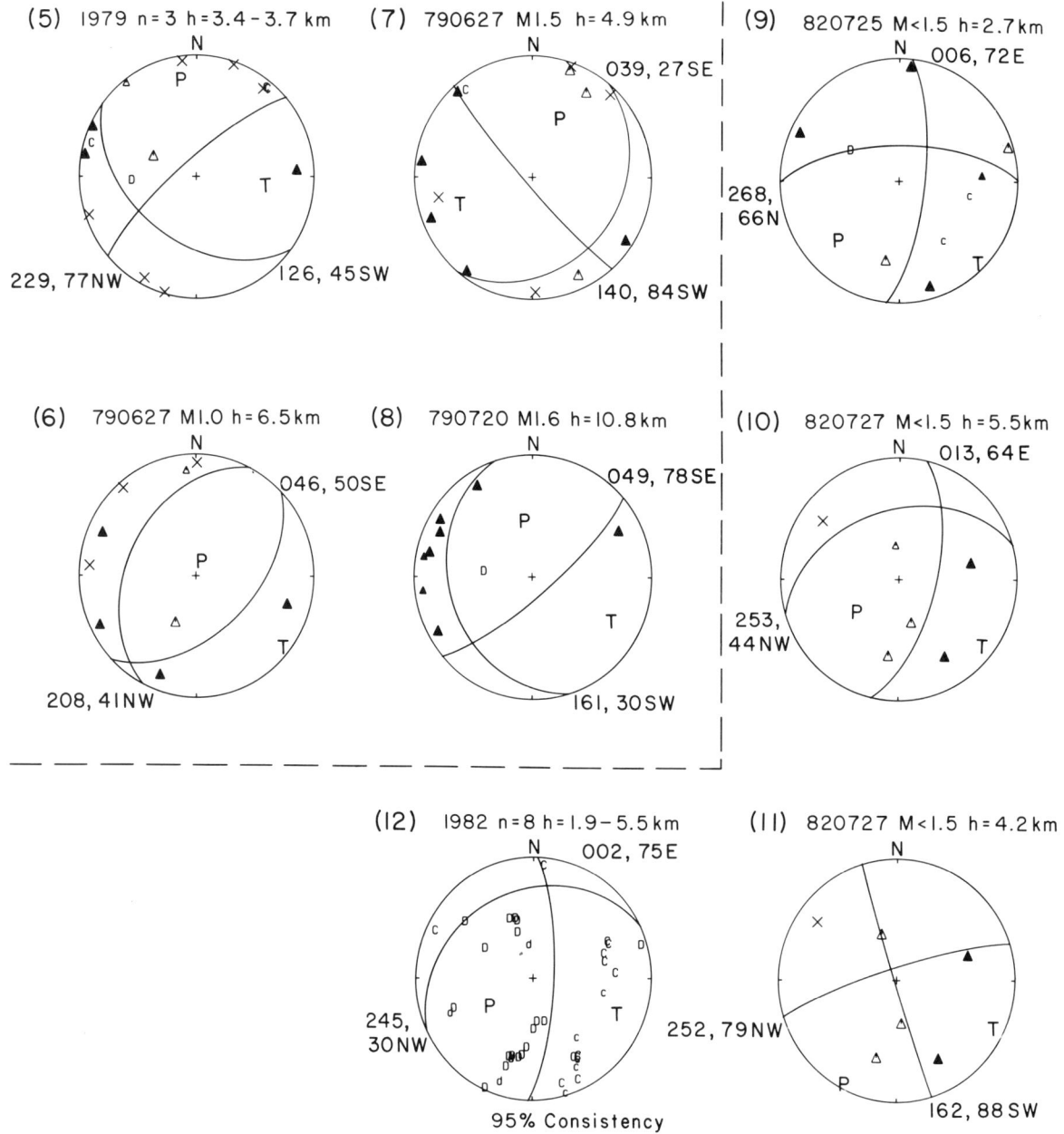

Figure 9. Fault-plane solutions (keyed to Figs. 7 and 8) determined in this study for earthquakes in the vicinity of the southern Wasatch fault zone. Symbols as in Figure 6. For solutions 5 through 11, nodal planes are based on SV/P amplitude-ratio inversion; for solution 12, P-wave first motions.

cisely located shocks from that sample, and solutions 9–11 as single-event, amplitude-ratio solutions for three randomly selected sub-events of solution 12.

The fault-plane solutions in Figure 9 display considerable variability in nodal-plane orientation, but remarkable consistency in T-axis orientation. Seven of the eight solutions have a T-axis with shallow plunge to the east-southeast or southeast. Normal faulting predominates, but there is substantial mixing of strike slip. In answer to our posed question, solutions 5, 6, and 7 do not have a low-angle, westward-dipping nodal plane correlative with the low-angle faulting shown in Figure 8. Solution 8, however, does have such a plane dipping 30°WSW, and is for an earthquake within or very close to the top of Precambrian basement. The nodal plane with smallest dip is one of solution 7 dipping 27°SE. For the swarm earthquakes, a common feature of solutions 9–12 is a northerly-trending nodal plane that dips fairly steeply eastward to near-vertical; the P-wave composite (solution 12) has an inferred nodal plane dipping 30°NW.

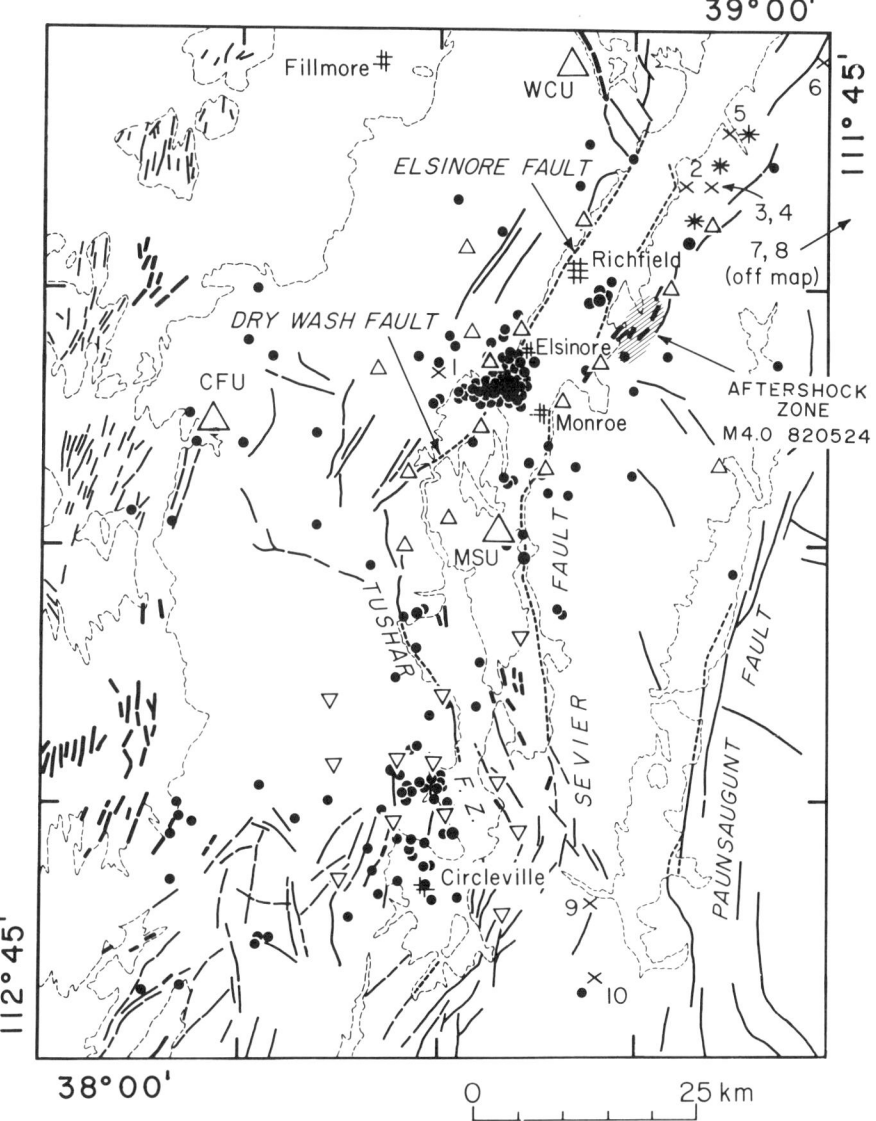

Figure 10. Summary map of earthquake field experiments in the Sevier Valley area during 1981 (1982 follow-up study within aftershock zone near Richfield). Base map as in Figure 2, but more detailed. Solid circles are epicenters (with mean epicentral precision <2 km); triangles, seismograph stations (larger ones, permanent stations; smaller ones, temporary—associated with two separate arrays represented by upright and inverted small triangles); X's and asterisks, sites of drill holes and quarry blasts, respectively, used for special investigation of upper-crustal structure. Cenozoic faulting shown by heavy lines—heaviest for late Quaternary fault scarps in unconsolidated alluvium (Anderson and Bucknam, 1979).

We return later to these results after presenting data for other localities. At this point, we observe the following. The summary information in Figure 8 indicates a random sampling of background seismicity—perhaps mostly reflecting slip on secondary faults. For the sampled seismicity, foci predominate above an inferred major detachment. Importantly, we can argue from our results that epicentral clustering in the vicinity of the southern Wasatch fault (Figs. 3 and 7) cannot be simply ascribed to seismic slip on the Wasatch fault itself without new substantiating data. Figures 7 and 8 show that: (1) earthquakes with epicenters along the trace of the Wasatch fault are not spatially associated with the fault at depth; and (2) earthquakes clustered west of the fault are not compatible with either down-dip planar projection of the fault or low-angle slip on a listric projection of the fault.

Sevier Valley

Field experiments in the Sevier Valley area (Fig. 10) during 1981 and 1982 have provided some of the highest-quality data of our current program of studies of the BR-CP transition. Special

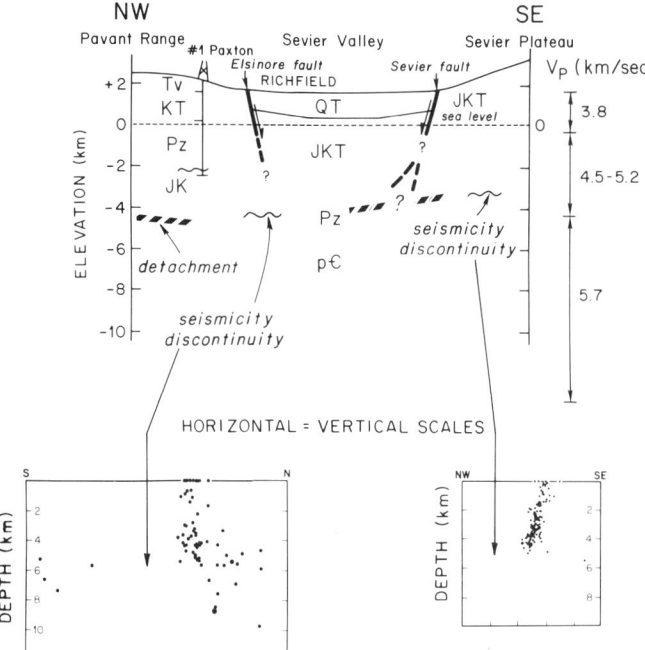

Figure 11. Schematic section across the Sevier Valley near Richfield, Utah, illustrating key results from local earthquake studies. Spatially discontinuous seismicity with depth appears to reflect local structural control by an inferred low-angle detachment (see text for discussion). Local P-wave velocity structures determined from nearby quarry blasts as refraction sources and by analysis of local earthquake data for multilayering (see Appendix). Standard abbreviations indicate geologic age of rocks; Tv = Tertiary volcanic rocks. Note that section in lower left is transverse to upper cross section. Datum for earthquake cross sections (below) is 1,500 m above sea level; transposition to upper section correspondingly adjusted. (See text regarding focal-depth precision.)

efforts were made (1) to refine upper-crustal velocity structure (see Appendix), (2) to achieve high spatial precision of hypocenters through strategic field recording and JHD processing, and (3) to determine numerous focal mechanisms using both P-wave first motions and SV/P amplitude inversion. In this section, we outline first-order results—with complete reporting left for a separate paper (D. R. Julander and W. J. Arabasz, in prep.; see also Julander, 1983).

Figure 10 gives an overview of our field experiment during the summer of 1981, which involved about five weeks of recording in the vicinity of Richfield (stations indicated by small upright triangles) and two weeks of subsequent recording to the south along the Tushar fault zone (stations indicated by small inverted triangles). Ten months later a detailed 8-day aftershock study was carried out following an earthquake of magnitude 4.0 on May 24, 1982, several kilometers southeast of Richfield (Fig. 10). Fortunately, the 1982 seismicity provided an opportunity to acquire high-quality earthquake data on the eastern side of the Sevier Valley—supplementary to abundant data acquired in 1981 on the western side.

Prominent features of the June–July 1981 earthquake sample shown in Figure 10 are the intense clustering of earthquakes (M ≤1.9) in the Elsinore area southwest of Richfield, and clustered but more diffuse earthquakes (M ≤1.2) near the southern end of the Tushar fault zone. The earthquake catalog corresponding to Figure 3 indicates persistent background seismicity in both of these areas. Prior to the 1981 field recording, there had been no earthquakes greater than magnitude 2 close to Elsinore since a pair of magnitude 4 shocks in 1972—inadequate to explain the observed clustering 9 years later (regarding effects of antecedent earthquakes on microseismicity see Arabasz and Robinson, 1976). In the southern area of epicentral clustering, the last shock greater than magnitude 2 was one of magnitude 3.5 in 1974.

At small scale, the earthquake clustering in the northern part of Figure 10 has a center roughly 4 km southwest of Elsinore. The clustering is positioned in an echelon gap between the northeast-trending Elsinore and Dry Wash faults and shows traverse east-west epicentral alignment. Cross sections of JHD-processed earthquake locations are shown in Figure 11 for both the 1981 earthquakes near Elsinore (66 JHD locations from a sample of 117 hypocenters) and the 1982 aftershocks southeast of Richfield (123 JHD locations from a sample of 242 hypocenters). For the first area, epicentral and focal-depth precision, respectively, are 0.80 (±0.41) km and 1.24 (±0.93) km; for the second area, 0.74 (±0.21) km and 1.21 (±9.4) km.

The cross sections of earthquake foci in Figure 11 indicate a discontinuity in the depth distribution of seismicity at about 5–6 km below datum in the vicinity of the Sevier Valley. Foci on the eastern side of the valley (lower right) abruptly terminate at this level; foci along the western side of the valley (lower left), plotted in a north-south plane, decrease markedly in number below this level. The schematic cross section in the upper part of Figure 11 shows that these discontinuities in seismicity coincide approximately with (1) a jump in P-wave velocity, and (2) an inferred low-angle detachment. The segment of a detachment indicated beneath the Pavant Range is from an interpretation of unpublished seismic-reflection data (R. B. Smith, 1983, oral communication) and is assumed to be correlative with a regional Mesozoic thrust identified by Allmendinger and others (1983, event E) from COCORP profiling to the west. The approximate depth and continuity of the inferred low-angle detachment beneath the Sevier Valley (Fig. 11) agrees with unpublished industry data. Without being able to display such information, we simply argue for the plausibility of our interpretation from the structural style documented for this general area by Smith and Bruhn (1984) and Standlee (1982).

Twenty-five single-event focal mechanisms, shown in Figure 12, were determined from the Sevier Valley data set using the amplitude-ratio technique. These include nine from the Elsinore area (solutions a–i; h = 3.3–7.4 km), ten from the aftershock area southeast of Richfield (solutions p–y, h ≤3.9 km), two from an area about 25 km south of Richfield (solutions n–o, h = 6.1–6.6 km), and four from the vicinity of the Tushar fault zone (solutions j–m, h = 5.5–9.5 km). The sampled focal mechanisms indicate a clear predominance of strike-slip faulting. Careful attempts were made to discriminate variations in focal mechanisms with depth

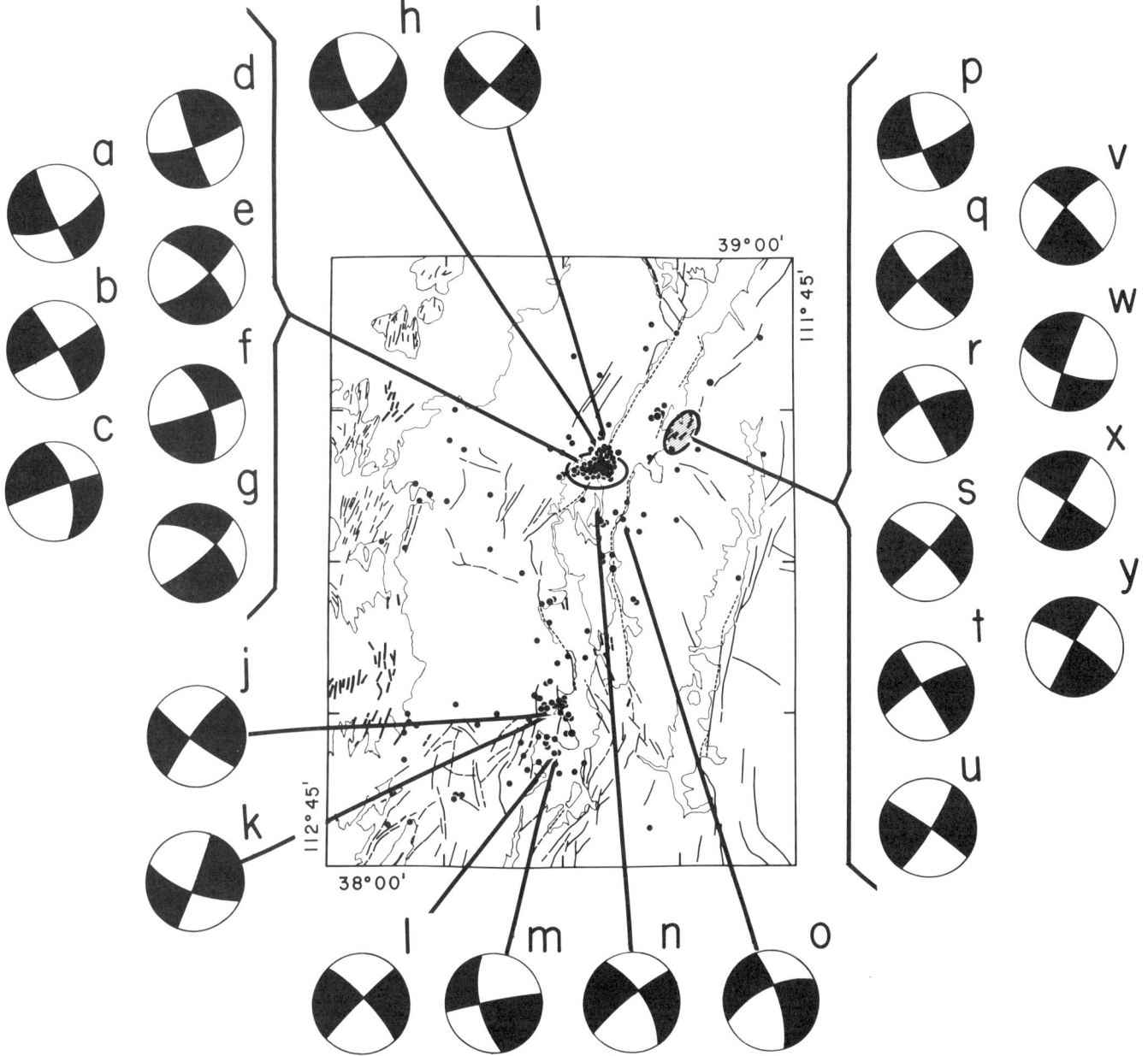

Figure 12. Summary of focal mechanisms (lower-hemisphere, compressional quadrants black) determined from 1981-1982 studies in the Sevier Valley area.

(Vetter and Ryall, 1983), but no evidence was found for any such depth dependence down to a sample depth of 9.5 km. Nor was evidence found for variation in focal mechanism above and below the inferred horizontal detachment beneath the Richfield area.

Figure 12 shows the mixture, in two separate source regions, of strike-slip mechanisms with opposing senses of slip. Compare the grouped solutions a–c with d–g, and p–u with v–y. In each respective group the left-hand column illustrates mechanisms with left-lateral slip on a northeasterly-trending nodal plane; the right-hand column, just the opposite. To demonstrate the reliability of these contrasting mechanisms, composite P-wave first-motion data are presented in Figure 13 for two populations of accurately located aftershocks from which solutions p–u and v–y were taken. One population (Fig. 13, lower left) consists of earthquakes having a clear compressional first motion at station GLN, the closest station; the second population (lower right), earthquakes with clear dilatational first motions at GLN. For the latter data, the pair of nodal planes plotted as solid lines are preferred to the dashed alternative pair on the basis of the corresponding single-event amplitude-ratio solutions (solutions v–y, Fig. 12).

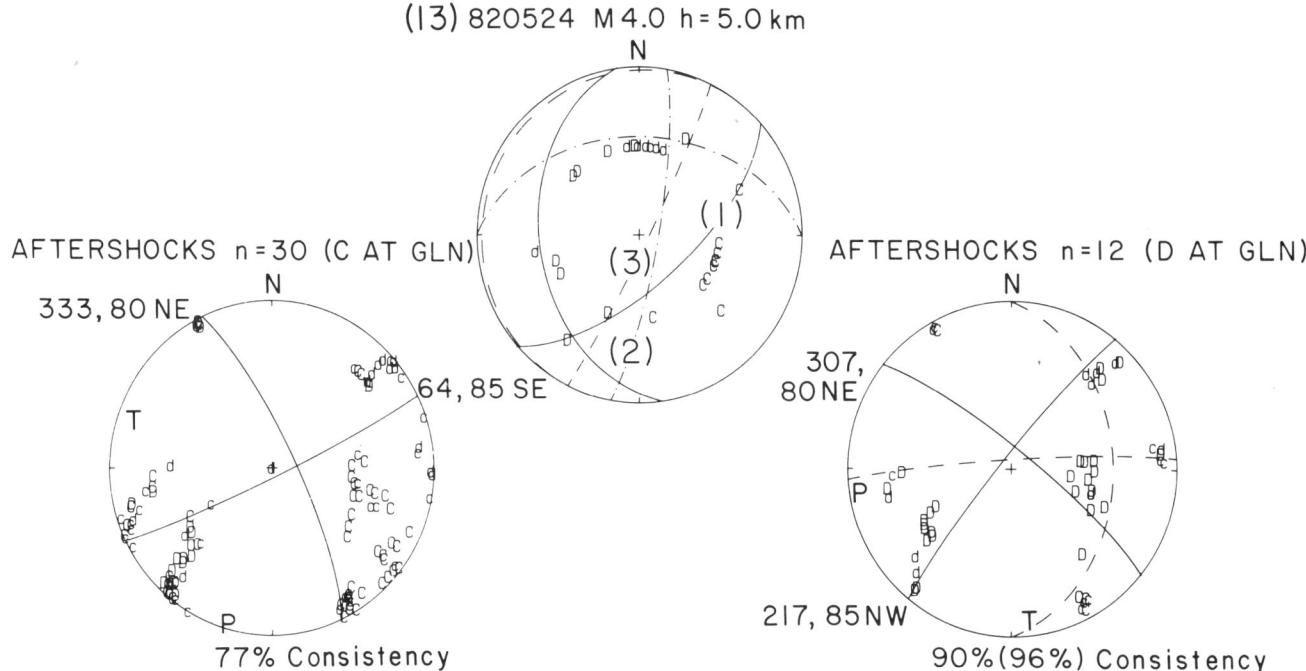

Figure 13. Fault-plane solutions for May 24, 1982, earthquake near Richfield, Utah (above), and following aftershocks (below). Symbols as in Figure 6. (See text for discussion.)

Possible explanations for the contrasting strike-slip mechanisms include: (1) observational error (which we argue against), (2) sequential episodes of opposite slip on the same or similar fault planes, (3) coherent block motion along specific planes resulting in slip vectors that do not coincide with directions of maximum resolved shear stress (Hill, 1982), and (4) local rotation of stress due to discontinuous strike-slip fault geometry (Segall and Pollard, 1980). We have no fully satisfactory explanation for the opposing strike-slip mechanisms, which are all for microearthquakes smaller than magnitude 1.5. On the basis of regional focal-mechanism data (summarized later) and fault-slip studies from the immediate vicinity (Anderson and Barnhard, 1984), the sampled earthquakes exhibiting right-lateral slip on northeast-striking nodal planes are considered to be anomalous and not reflective of a regional stress field.

The 1982 earthquake sequence near Richfield poses yet another enigma. Reliable first-motion data for the mainshock cannot be simply reconciled with either the planar distribution or sense of slip of aftershocks (Fig. 13). The aftershock foci (Fig. 11; Julander, 1983) define a planar zone that strikes N60°–70°E and dips 75°NW; within this plane the foci outline an elongate patch 2–4 km wide that plunges to the north. The surface trace of the planar zone coincides roughly with northeast-trending, down-to-the-west fault scarps of late Quaternary age mapped by Anderson and Bucknam (1979) along a branch of the Sevier fault (see Fig. 10).

First-motion data for the mainshock, recorded by the regional seismic network at distances of 24–280 km, are plotted in Figure 13 (center) as a function of focal-sphere position computed by J. C. Pechmann (1984, written communication). A focal depth of 5.0 km was assumed from the aftershock distribution and was preferred to a free-depth solution of 2.0 km, which is unrealistically shallow for a magnitude 4 earthquake. The mainshock data require some component of normal slip, but nodal-plane constraint is poor—as indicated by the validity of three alternative pairs. Slip vectors for these alternative solutions (see solution 13, Table 1) have rakes ranging from 18° to 137°. The first-motion data do not allow interpretation of a fault plane dipping steeply to the northwest, as suggested by the surface geology and aftershock foci, nor do they allow the predominance of strike slip on a near-vertical plane as displayed by the aftershocks. Each of the three alternative fault-plane solutions for the mainshock does, however, have an east-southeast-trending T-axis compatible with the majority of strike-slip aftershocks. Given solution 13(3) (Fig. 13, Table 1), one might argue that pure normal slip occurred during the mainshock on a near-horizontal plane dipping 5°NW—followed by aftershock readjustment on a high-angle plane within the upper plate. While such a speculative interpretation agrees neatly with our inference of a low-angle detachment, we do not believe that seismic slip on a low-angle plane would lead to a rupture zone (including boundary) completely devoid of aftershocks. The kinematics of faulting associated with this unusual mainshock-aftershock sequence remains to be deciphered. Our suspicion, based on the plunging elongate patch of aftershock foci, is that a buried fault intersection may have been involved.

The significance of our observation of abundant strike-slip focal mechanisms in the Sevier Valley area was initially uncertain. Obvious objections included (1) the assumed morphotectonic dominance of normal faulting, (2) the apparent absence of any geological evidence for local strike-slip deformation, and (3) antecedent focal mechanisms, such as that for a magnitude 5.2 earthquake in 1967 near Marysvale (Table 1, solution 15) for which Smith and Sbar (1974) had interpreted nearly pure dip slip on a near-vertical, northerly-trending fault plane. (We disregard a composite focal mechanism determined by Sbar and others [1972] for the Marysvale area, judging it to be unreliably based on graphical earthquake locations and too broadly scattered foci.) Subsequent geological studies by Anderson and Barnhard (1984; also R. E. Anderson and T. P. Barnhard, in prep.) have uncovered abundant evidence of young strike-slip faulting in the Sevier Valley area, including late Cenozoic horizontal displacements that may approach 10 km. Range-flanking alluvium typically obscures major faults bounding the Sevier Valley, and it now appears that monoclinal flexuring and oblique slip have been important in producing structural relief (Anderson and Barnhard, 1984). H. J. Patton (1985, written communication) has re-evaluated the 1967 Marysvale mainshock using surface-wave inversion and computes a mainshock mechanism involving a slip-vector rake of 144° on a nodal plane striking N10°W and dipping 74°E; the computed focal depth is 14 km. Thus, oblique to strike-slip displacements appear not to be restricted to small-magnitude earthquakes in the Sevier Valley area, nor to be restricted to the shallow uppermost crust. Implications of these new results are only beginning to be pursued.

Eastern Wasatch Plateau

We earlier noted the association of intense seismicity in the eastern Wasatch Plateau with underground coal mining (see "Regional Seismicity"). Here we briefly consider some observations from a 1979 reconnaissance study in that area carried out by McKee and Arabasz (1982) that were reported only in unpublished form (McKee, 1982). The observations add to our perspective of seismic deformation within the BR-CP transition but should, however, be considered provisional. A more extensive portable-array study just carried out in the same area by one of us (WJA) during the summer of 1984 will provide more rigorous detail.

Figure 14 illustrates the following results from the 1979 recording in the East Mountain area of the eastern Wasatch Plateau: (1) intense spatial clustering of seismicity (h <4 km) both at and beneath levels of active underground coal mining; (2) the occurrence of microearthquakes 6–16 km deep beneath the shallow mining-related seismicity; and (3) a composite fault-plane solution for 21 earthquakes in the 1.5 to 2.8 km depth range that indicates reverse faulting and northwest-trending nodal planes—contrasting with north-to-northeast–trending, post-Eocene normal faulting mapped at the surface.

Smith and others (1974) had earlier observed coal-mining-

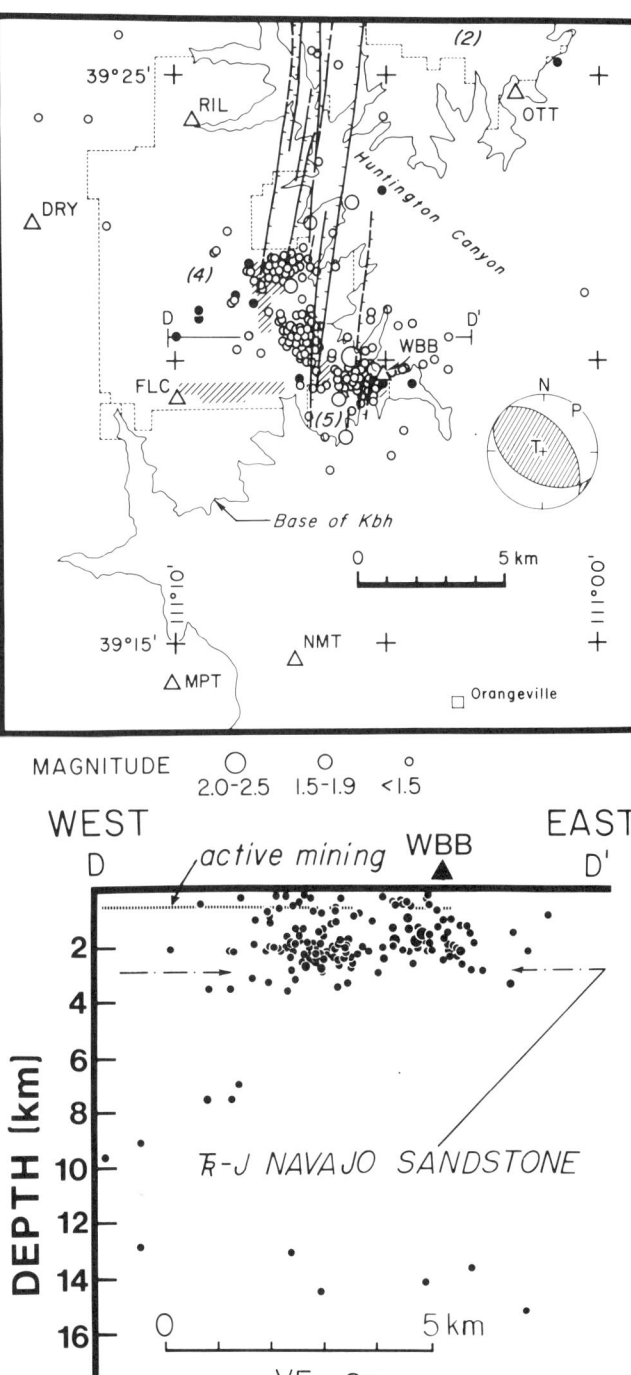

Figure 14. Results from reconnaissance earthquake studies in the eastern Wasatch Plateau (adapted from McKee, 1982). Epicenter map (above) is for the period August 8–14, 1979. Triangles indicate temporary seismograph sites within bounds of the figure; open and closed circles, foci shallower and deeper, respectively, than 3.9 km below datum level of 2,750 m above sea level; hachured regions, areas of active underground coal mining during study period; light dashed lines, boundaries of coal-mining properties; Kbh, Cretaceous (coal-bearing) Blackhawk Formation; heavy lines, faults; beachball (lower-hemisphere, compressional quadrant shaded), composite fault-plane solution (from McKee, 1982). Cross section (below) illustrates location of microearthquake foci with respect to level of active mining and position of Tr-J Navajo Sandstone.

induced seismicity down to 2 km beneath a mine 60 km to the east in the Book Cliffs area. Apart from seismic strain release within a few hundred meters of mine workings, induced seismicity at distances of kilometers is unusual, but perhaps also occurs beneath some coal mines in Poland (Gibowicz and others, 1981). In such cases, triggering of tectonic stress is presumed to be a factor.

Wong (1984) has investigated two-dimensional finite-element modeling of a typical Wasatch Plateau–Book Cliffs coal mine to evaluate in-situ stress induced by mining. Stress redistribution accompanying underground coal excavation beneath cliff topography is shown to produce changes in vertical stress on the order of a few bars or less at depths of 1 to 3 km below mine workings—sufficient to trigger slip on tectonically prestressed reverse faults. The presence of a low-angle detachment beneath the eastern Wasatch Plateau, discussed earlier in connection with the geologic framework of the study area, could also conceivably exert an important influence on the depth distribution of sub-mine earthquakes. Following the interpretation of Standlee (1982, Fig. 14) that such a detachment may lie within incompetent Jurassic strata overlying the Navajo sandstone, we have plotted the depth of the latter in Figure 14 and note significant correlation with a lower bound to the clustered shallow seismicity.

If a shallow low-angle detachment indeed underlies the eastern Wasatch Plateau, it could have a critical bearing on the mechanics and maximum size of induced shallow earthquakes—as well as on our understanding of crustal faulting and seismogenesis associated with prominent late Quaternary fault scarps along the nearby Joes Valley fault zone (Fig. 2; Anderson and Miller, 1980). Upper-crustal seismicity may similarly be influenced by related low-angle detachments both in the Sevier Valley area to the west and in the eastern Wasatch Plateau. Upper-plate stress orientation, however, apparently differs between the two locales, a point we return to later.

SEISMIC SLIP AND SUBSURFACE FAULT GEOMETRY

Observations relevant to subsurface fault geometry along the eastern margin of the BR Province are (1) of considerable local interest, because of seismotectonic issues raised by the interpreted widespread presence of listric normal faults and low-angle detachments (see Smith and Bruhn, 1984), and (2) of general interest, because of current attention to the kinematics of extensional tectonics (e.g., Wernicke and Burchfiel, 1982; Brun and Choukroune, 1983; Jackson and McKenzie, 1983). Here we simply ask the question, What do earthquake observations imply about subsurface fault geometry in the study area? Fault-plane solutions and spatial patterns of earthquake foci are the basic sources of information.

Fault-plane solutions in central and southwest Utah have been used to argue for the predominance of seismic slip on fault segments with moderate to high-angle dip—with weak evidence

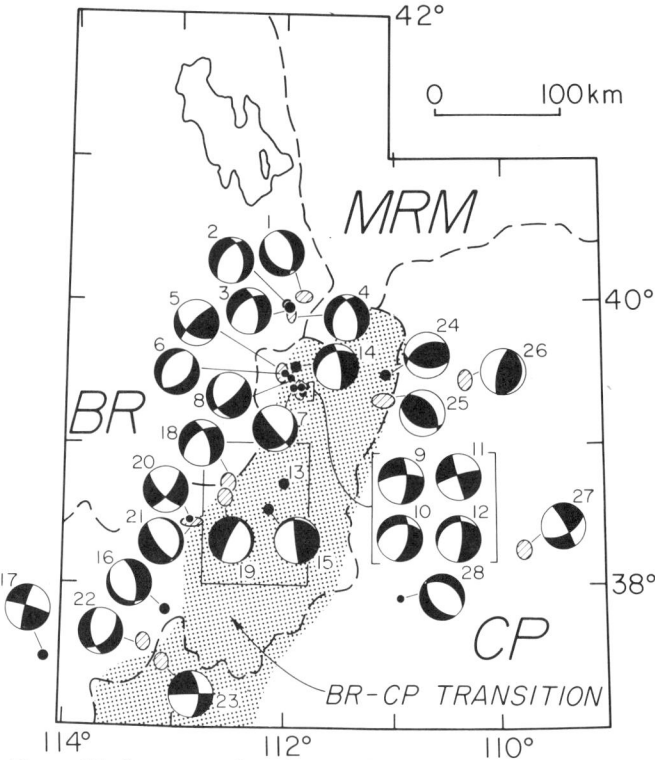

Figure 15. Summary of earthquake focal mechanisms (lower-hemisphere, compressional quadrants black) across the Basin and Range–Colorado Plateaus transition. (Alternative outlines and abbreviations as in Figure 1.) Solution numbers keyed to Table 1. Large and small dots show locations of single-event solutions (respectively for M ≥3.5 and M <3.5), and hachured zones show sample areas for composite solutions. Location 13 is for the M4.0 earthquake near Richfield in 1982 (Fig. 13). Box in central part of transition outlines area of additional 25 fault-plane solutions shown in Figure 12.

for seismic slip on low-angle (<30°) planes (Arabasz, 1981, 1983; McKee and Arabasz, 1982). Zoback (1983) reached similar conclusions reviewing fault-plane solutions for northern Utah, noting that a sample of 22 nodal planes for 11 solutions has a range of dips between 20° and 80°, a mean dip of 49°, and median dip of 49° to 54°. We turn to a systematic summary of fault-plane solutions throughout central and southern Utah for information on subsurface fault dip. The data will also provide a basis in a later section for investigating implications for stress orientation.

Figure 15 summarizes available fault-plane solutions between lat. 37°N and 40°N in central and southern Utah. Counting the 25 solutions of Figure 12, the sample consists of 53 solutions, 39 of which come from our program of studies of the BR-CP transition. Parameters for the solutions in Figure 15 are given in Table 1. For this same area, Smith and Lindh's (1978) compilation had only eight fault-plane solutions, three of which we have not included. Two of these (their solutions 7 and 8) are composite solutions from Sbar and others (1972), which we consider unreliable because too broadly scattered earthquakes were

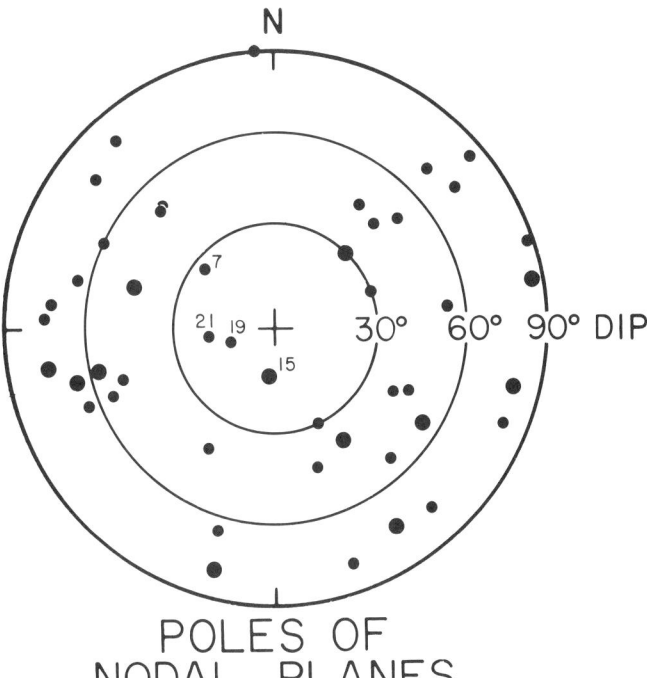

Figure 16. Stereographic plot (lower-hemisphere, equal-area) of the poles of nodal planes for all fault-plane solutions in Figure 15 excluding solutions 24–27. (Solutions indexed to Fig. 12 also excluded.) Isocontours indicate values of dip angle for planes corresponding to the plotted poles. Points corresponding to nodal planes with less than 30° dip are numbered (keyed to Table 1). Large and small circles indicate earthquake size ≥M3.5 and <M3.5, respectively.

grouped to form the composites, and the focal-sphere data were based on graphical earthquake locations. The third (their solution 71) is ascribed to unpublished data but appears to be an erroneous duplicate of a nearby single-event focal mechanism.

The strike-slip solutions of Figure 12 are uninformative regarding seismic slip on low-angle faults because they all involve nodal planes with high-angle to near-vertical dip. Excluding those solutions, and also excluding compressional solutions 24–27 near or within the interior of the Colorado Plateau, let us consider the sample of nodal planes for the remaining focal mechanisms in Figure 15, which are predominantly extensional. Figure 16, a composite plot of the poles of all the nodal planes from those solutions, gives an effective overview of fault dips. Because the true fault plane cannot confidently be selected from each orthogonal pair, both nodal planes are plotted. Given the small-to-moderate size of the corresponding earthquakes and great uncertainty about the association of seismicity and structure in this region, it is arguable whether a fault plane can be preferentially chosen in any of the cases.

Figure 16 shows only four cases where one of the potential seismic-slip planes dips less than 30°. A fifth case is for one of the alternative nodal planes for the 1982 earthquake near Richfield (Fig. 13), not plotted. The low-angle nodal plane for solution 15 is from Smith and Sbar's (1974) focal mechanism for the magnitude 5.2 earthquake near Marysvale in 1967, which we described earlier as open to question. Taken as a group, the nodal planes plotted in Figure 16 have a range of dips between 14° and 90°, a mean dip of 57° (±20°), and a median dip of 55° (bearing in mind that only half the data points can be independent).

We do not argue that all low-angle nodal planes for the fault-plane solutions in the region are unreliable. For example, the nodal plane dipping 5°NW in solution 13 (3) (Fig. 13) must be considered a real possibility, and a similar low-angle nodal plane dipping 10°–22° has recently been reliably determined by Pechmann and Thorbjarnardottir (1984) for a magnitude 4.3 earthquake near Salt Lake City in 1983. The issue is whether there is any independent evidence that seismic slip occurred on the low-angle plane, rather than on the steeper complementary nodal plane. In neither of the above cases was there any suggestion of aftershock clustering on a low-angle plane.

Crone and Harding (1984) have interpreted a direct connection in western Utah between displacements on high-angle normal faults and displacements on the subjacent Sevier Desert detachment, but it remains moot whether slip on the detachment can occur seismically. It is commonly noted, for example, that for maximum principal stress oriented vertically (normal faulting), resolved shear stress on low-angle faults should be inadequate for frictional sliding (see Jaeger and Cook, 1979, p. 65), and special rheologic conditions must be adduced for such sliding to be feasible. We are aware of no instance in which clustered earthquake foci and corroborating focal mechanisms support the depiction of a continuous slip surface in the form either of a downward-flattening (listric) normal fault or a planar low-angle normal fault.

Our studies throughout the BR-CP transition confirm the predominance of seismic slip on fault segments of moderate (>30°) to high-angle dip—at least for small- to moderate-sized earthquakes—based on both fault-plane solutions and hypocentral distributions. In the Sevier Valley where upper-crustal faulting is inferred to merge with a subjacent low-angle detachment (Fig. 11, lower right), earthquake foci indicate planar normal faulting on a high-angle plane and sharp discordance with the detachment rather than downward flattening into it. For background seismicity downdip of the Wasatch fault near Goshen Valley (Fig. 5) and along the southwest flank of the Gunnison Plateau (Fig. 8), foci are incompatible with seismic slip on a listric projection of the Wasatch fault.

The southern Wasatch fault is one locality where there is reliable independent evidence for low-angle normal faulting in the subsurface, but seismic slip on fault segments of moderate to high-angle dip was found to predominate (Figs. 8 and 9). Given the widespread presence of listric and low-angle normal faults in the study area, seismic slip on steeper fault segments might be variously explained as associated with the upper portion of downward-flattening faults, related antithetic faults, or secondary block-interior faults (as opposed to block-boundary faults of greater scale). Perhaps only the experience of future large earthquakes will resolve whether the low-angle fault segments can be engaged seismically.

CORRELATION OF SEISMICITY WITH GEOLOGIC STRUCTURE

To say that the correlation of diffuse seismicity with geologic structure in the Utah region is problematical has become trite, but remains a fact. Problems include: (1) uncertain subsurface structure, which we have seen is more complex than apparent from the subsurface geology; (2) observations of discordance between surface fault patterns and seismic fault slip at depth (Arabasz and others, 1981; Zoback, 1983); (3) a paucity of historic surface faulting; and (4) inadequate focal-depth resolution from regional seismic monitoring. Even with good hypocentral precision from our portable-array studies, it is evident that adequately concentrated seismicity and abundant single-event focal mechanisms are required to unravel the association of seismicity with structure in the absence of a large earthquake.

In attempting to explain the diffuse nature of background seismicity in central Utah, McKee and Arabasz (1982) considered various possibilities, including: (1) poor hypocentral resolution; (2) complex subsurface structure; (3) depth-varying stress orientations, which might result in superposed epicentral patterns from differently oriented planes of weakness at different depths; and (4) variations in seismicity patterns with stress level or temporal stage of a large earthquake cycle (e.g., Scholz, 1968; Kanamori, 1981). To this list of possibilities we can add: (5) laterally inhomogeneous stress fields, such as off-fault stresses (Das and Scholz, 1981); (6) preferential seismic slip on small block-interior faults rather than on major block-boundary faults (Hill, 1982), perhaps resulting from structural control on permeability (Leith and others, 1981); and (7) geometric irregularity caused by barriers and asperities (see Aki, 1984).

There has been only one instance of historic surface faulting in the Utah region—a vertical displacement of 0.5 m associated with the magnitude 6.6 Hansel Valley earthquake of 1934 (see Arabasz and others, 1980). The occurrence of seven other historical earthquakes of magnitude 6.0 to 6.6 without surface rupture (Arabasz and others, 1979) suggests a relatively high threshold for surface faulting in the region. Bucknam and others (1980) summarize data on documented surface faulting in the BR province and note that all historic earthquakes in the Great Basin of magnitude 6.3 or greater (7 cases) have produced surface rupture. The two smallest earthquakes in their tabulation (magnitude 5.6 and 6.3), had maximum displacements of 0.2 m and 0.1 m, respectively; earthquakes of magnitude 6.8 or less (5 cases) all had a maximum displacement less than one meter.

Rigorous efforts were not made to search for evidence of surface faulting immediately following many of Utah's larger historical earthquakes, say of magnitude 5 or greater, so the threshold for small surface displacements up to a few tenths of a meter is debatable. Moreover, the minimum magnitude for surface rupture depends on variable parameters of an earthquake source, including focal depth, fault geometry, stress drop, seismic moment, rupture-propagation dynamics, and the like. In the most recent case of a thoroughly studied earthquake of significant size in the Utah region, the magnitude 6.0 Pocatello Valley (Idaho-Utah border) earthquake of 1975, no evidence was found for tectonic surface faulting, but observations were made of nontectonic surface cracking in frozen snow cover in the alluviated valley, and up to 13 cm of localized subsidence may have been coseismic (Arabasz and others, 1981).

The relatively high threshold of surface faulting and observations of discordance between surface fault patterns and seismic slip at depth have profound implications. First, the location and geometry of potential sources for local earthquakes below the threshold of surface faulting (approximately magnitude 6 to 6½) cannot be confidently assessed by knowledge of the surface geology alone. Where subsurface structure is complex, moderate sized earthquakes may occur on "blind" subsurface structures that have no direct surface expression. Second, the nucleation of large surface-faulting earthquakes is obscured.

Crustal structure along the eastern BR province is now believed to involve vertically stacked plates separated by low-angle detachments resulting from pre-Neogene thrustbelt structure and/or Neogene extension (Allmendinger and others, 1983; Smith and Bruhn, 1984; among others). Results of our portable-array studies lead us to the following working hypothesis to explain observations of diffuse background seismicity in the region. We suggest that background seismicity is fundamentally controlled by variable mechanical behavior and internal structure of individual horizontal plates within the seismogenic upper crust. Diffuse epicentral patterns result from the superposition of seismicity occurring within individual plates, and also perhaps from favorable conditions for block-interior rather than block-boundary microseismic slip. The depth distribution of earthquakes in Figures 8, 11, and 14 suggests that background seismicity predominates within the uppermost plate.

We believe that the preponderance of background earthquakes of small to moderate size in central Utah, between about the latitudes of Nephi on the north and Elsinore on the south, occur shallower than 6–7 km depth above low-angle detachments. Patton (1985) has recently estimated a focal depth, based on surface-wave inversion, of 4 km for the magnitude 4.4 earthquake in Juab Valley in 1963 (Fig. 3, epicenter northwest of station LVU). At the same time it is clear that background earthquakes are occurring deeper (7–15 km), albeit less frequently, beneath inferred low-angle detachments in this same region of central Utah (Figs. 8, 11, and 14).

To reinforce the credibility of our idea that low-angle detachments may exert a fundamental influence on background seismicity, we show results in Figure 17 from a special study of earthquake swarm activity ($M_L \leq 4.7$) in October 1982 near Soda Springs, Idaho (42.6°N, 111.4°W). We have superposed JHD-processed earthquake foci from Richins and others (1983) upon a geological cross section generalized from Dixon (1982), based upon surface mapping and seismic-reflection data. The cross section shows the superposition of Neogene-Quaternary normal faulting upon Mesozoic-Paleogene thrustbelt structure—as in central Utah. Figure 17 shows first that seismicity is not

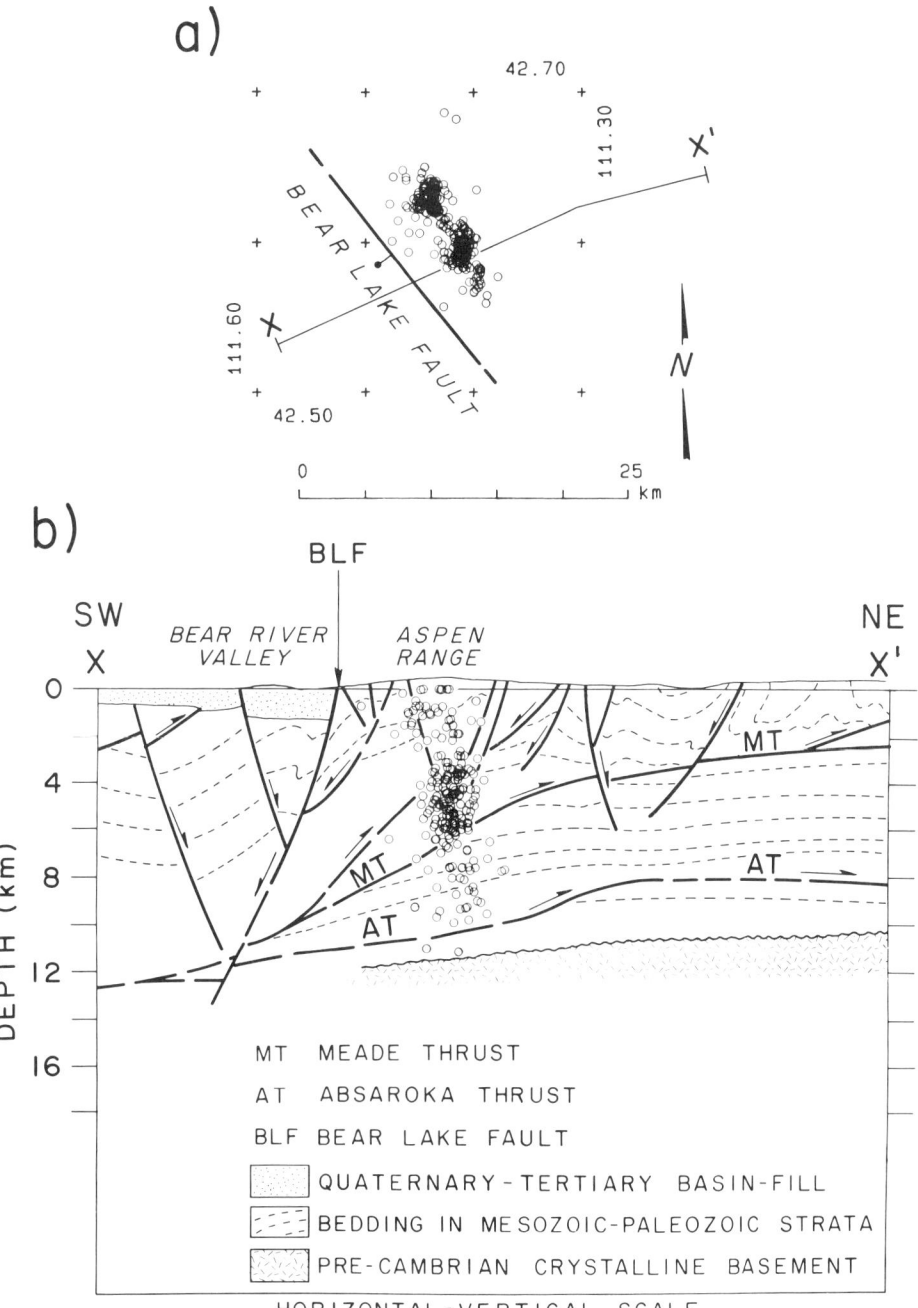

Figure 17. Association of swarm seismicity in late 1982 near Soda Springs, Idaho, with geologic structure. Earthquakes located by Richins and others (1983) are shown (a) in map view and (b) superposed upon a geological cross section (generalized after Dixon, 1982). Datum for the cross section is 1,700 m above sea level. Earthquake foci have a mean epicentral precision of 0.50 (±0.21) km and a mean focal-depth precision of 0.98 (±0.57) km.

simply associated with late Cenozoic basin-range faulting notably represented by the Bear Lake fault. The earthquake foci cluster within a northwest-trending, near-vertical zone *within* the Aspen Range block and not along its boundary with the Bear River Valley graben. Second, sharp changes in the frequency distribution of focal depths coincide with low-angle discontinuities. Of the 219 foci plotted in Figure 17, 84% lie above the Meade thrust (at 7 km depth), 15% within the underlying Absaroka thrust plate (between 7 and 10 km depth), and all lie above the interface between sedimentary cover and Precambrian crystalline basement (at 11–12 km depth). Despite the generally assumed tectonic predominance of basin-range normal faulting, focal

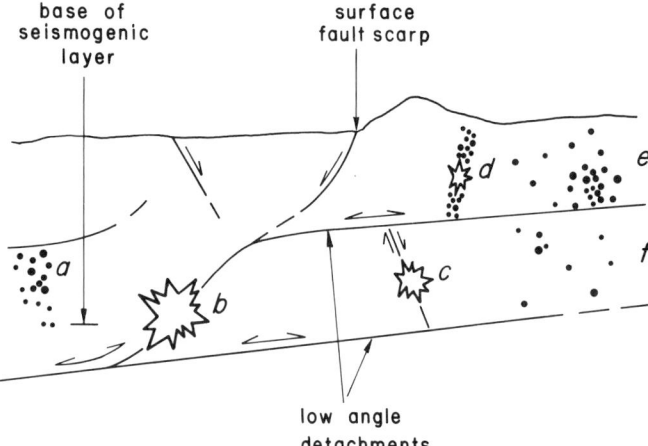

Figure 18. Schematic geologic cross section of the upper crust illustrating complex association of seismicity with geologic structure in the study area. Starbursts indicate foci of moderate-to-large earthquakes; small circles, microseismicity; lines in subsurface, faults. Arrows indicate sense of slip on faults; two-directional arrows, extensional backsliding on pre-existing low-angle faults possibly formed as thrust faults. Letters identify examples referred to in text. Base of seismogenic layer is approximately 10–15 km depth.

mechanisms both above and below the Meade thrust fault show a predominance of strike-slip faulting on northwest-trending, steeply-dipping fault planes—with no indication of seismic slip on a low-angle plane; some mixing of normal-faulting mechanisms was observed below the Meade thrust (Richins and others, 1983; and University of Utah unpublished data).

An important implication of Figure 17 (and also Fig. 11) is that the frequency distribution of focal depths within the upper crust may locally be controlled by structure—and not necessarily reflect rheological changes with depth (Smith and Bruhn, 1984). Maximum focal depths, on average, will expectedly reflect the frictional to quasi-plastic transition (Sibson, 1982). Our data suggest caution, however, in using other details of focal-depth frequency distributions as empirical evidence for the validity of specific rheological parameters.

Another important implication of our data is that precisely resolving the focal depth of background seismicity may be crucial for effective surveillance in the Intermountain seismic belt. Because large earthquakes are expected to nucleate at the base of the seismogenic layer (Sibson, 1982; Das and Scholz, 1983), at about 10–15 km depth in the Utah region (Smith and Bruhn, 1984), the pattern of background seismicity at this depth may in some cases be masked or blurred by greater flux of small- to moderate-sized earthquakes within an overlying plate. Such an hypothesis requires "linkage" or some rupture pathway between the deep nucleation points of large earthquakes and existing surface fault scarps. The nucleation depths of the two largest historical earthquakes in the ISB, the 1959 Hebgen Lake, Montana, earthquake (M_s7.5) and the 1983 Borah Peak, Idaho, earthquake (M_s7.3) were both at approximately 15 km (Doser, 1985).

Figure 18 schematically shows some aspects of the working hypothesis proposed here as relating diffuse seismicity to vertically stacked plates in the upper crust. (The hypothesis applies not only to the study area but to much of the Intermountain seismic belt.) These aspects, by no means exhaustive, include: (a) local predominance of background seismicity within a lower plate; (b) nucleation of a large normal-faulting earthquake near the base of the seismogenic layer, on an old thrust ramp, and with linkage or an established rupture pathway to a major surface fault; (c) occurrence of a moderate-sized earthquake within a lower plate—manifesting structural discordance with surficial geology, and with surface rupture inhibited by no established linkage to a shallow structure; (d) occurrence of a moderate-sized earthquake and aftershocks on a secondary rupture where an underlying detachment restricts deformation to the upper plate; (e) diffuse block-interior microseismicity predominating within an upper plate—perhaps responding to extension enhanced by gravitational backsliding on an underlying detachment; and (f) diffuse block-interior microseismicity within a lower plate where frequency of occurrence is markedly lower than in the overlying plate.

In the case of location b in Figure 18, note that focal-depth resolution would be critical to discriminate background earthquakes associated with such a deeper nucleation zone from background seismicity associated with an overlying shallow structure. The depicted nucleation of macroseismic slip along an old thrust ramp is arbitrary—but represents one way to get linkage between deep and shallow structure where low-angle detachments are present. The 1983 Borah Peak, Idaho, earthquake (M_s = 7.3) has strengthened the belief that large basin-range type earthquakes may typically involve rupture on moderately dipping planar faults from mid-crustal nucleation depth to surface (e.g., Smith and others, 1984). It remains unclear, however, how such penetrative planar faulting can be reconciled with subsurface structure such as that shown in Figure 8 for the vicinity of the southern Wasatch fault where a Holocene surface displacement of 2.5 m has occurred (Schwartz and Coppersmith, 1984). The possibility of initiating seismic slip on major low-angle detachments (Crone and Harding, 1984) remains an uncertain "wild card." Another key issue, and one beyond the scope of this paper, is the interaction between crustal extension at depth and that manifested in the uppermost crust—particularly in the form of the shallow background seismicity that we observe. We refer the reader to Kligfield and others (1984) and Anderson and others (1983) for relevant discussion.

REGIONAL STRESS ACROSS THE BR-CP TRANSITION

Zoback and Zoback (1980) provide a broad overview of regional stress orientation in the vicinity of the BR-CP transition, and the stress state within and across the transition has been the object of idealized analysis by McGarr (1982). A primary purpose of this section is to use our fault-plane solutions to expand the observational data base relevant to the orientation and relative magnitude of stress across the BR-CP transition. There is substantial information pertinent to stress state across the eastern margin of the BR province between about lat. 40°N and 42°N, where converging evidence implies an east-west to west-southwest–east-northeast (N75°E) direction of least principal stress/strain (see review by Zoback, 1983). Between lat. 37°N and 40°, however, stress-indicator data have been notably more sparse, and a small number of fault-plane solutions in that area compiled by Smith and Lindh (1978; see our earlier discussion of "Seismic Slip and Subsurface Fault Geometry") formed most of the observational base there for Zoback and Zoback (1980; see also Sbar, 1982).

The nature of the stress field within the BR-CP transition is of considerable interest because it must reflect a change from west-northwest–east-southeast horizontal extension, within the BR province, to west-northwest–east-southeast horizontal compression, within the interior of the CP province (McGarr, 1982; see heavy arrows in Fig. 19). Fault-plane solutions currently offer the most feasible way of mapping and discerning spatial variations of stress orientation in the study area. This can be done by assuming a correspondence between *average* directions of near-horizontal P- or T-axes and principal stress/strain directions (Zoback and Zoback, 1980), and adopting the common assumption that one of the principal stress directions is vertical (Zoback and Zoback, 1980; McGarr, 1982).

In the map of Figure 19 we have plotted the horizontal projections of P- and/or T-axes (where appropriate) from the fault-plane solutions compiled in Figures 12 and 15, together with average regional directions of principal horizontal extension (or compression) inferred by Zoback (1983) and Zoback and Zoback (1980). For the fault-plane solutions, average orientations have been plotted in five cases for groups of similar, spatially clustered solutions. (P- and T-axes were not plotted for the minority group of eight strike-slip solutions of Fig. 12 that were earlier described as anomalous, nor for solution 15 corresponding to the 1967 Marysvale earthquake.) Also plotted in Figure 19 is the boundary between a low-heat-flow thermal interior of the CP and a high-heat-flow periphery to the west (Bodell and Chapman, 1982; Powell and others, 1983), potentially related to the boundary of a distinct stress province inferred by Zoback and Zoback (1980) for the interior of the CP.

Allowing for differing interpretations of the boundaries of the BR-CP transition, we observe the following in Figure 19. The easternmost part of the BR province and the western part of the BR-CP transition are characterized by mixed-mode faulting, including normal slip, strike slip, and oblique slip. Reverse-faulting mechanisms appear in the northeastern corner of the transition. There is a general pattern in the orientation of inferred horizontal principal stress directions: within the transition, minimum horizontal principal stress tends to be perpendicular to the BR-CP boundary, and maximum horizontal principal stress tends to be parallel to the boundary. East of the transition, the few available solutions indicate an exactly opposite pattern.

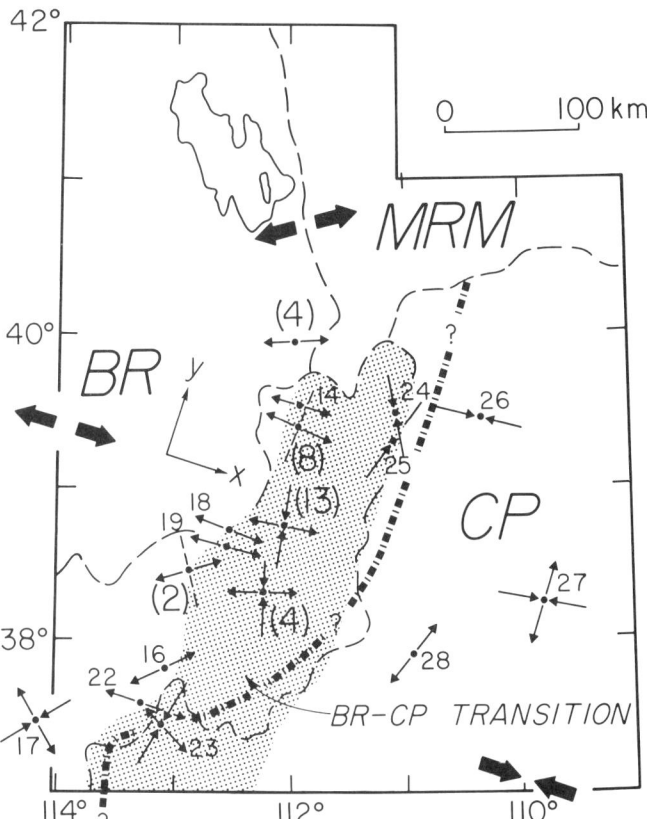

Figure 19. Schematic summary of stress-orientation data for the Basin and Range–Colorado Plateaus transition. Large arrows indicate regional directions of least (outward-directed) or greatest (inward-directed) principal horizontal compressive stress (from Zoback and Zoback, 1980; Zoback, 1983). Small arrows indicate inferred stress-orientation directions from P- and T-axes of fault-plane solutions shown in Figures 12 and 15: outward-directed arrows for normal-faulting mechanisms, inward-directed arrows for thrust or reverse-faulting mechanisms, a combination of inward and outward-directed arrows for strike-slip faulting, and outward-directed arrows with perpendicular dashed lines for mixed normal and strike-slip faulting. Small numbers identify data points keyed to Figure 15 and Table 1; large numbers in parentheses, the total number of other fault-plane solutions combined to produce an average orientation. The heavy patterned line (queried where uncertain) marks the limits of a low-heat-flow thermal interior of the Colorado Plateaus surrounded by a high-heat-flow periphery (after Bodell and Chapman, 1982, as modified by Powell and others, 1983).

For convenience, let us denote the three principal stresses as S_i, where the subscripts may either be numerical to denote relative magnitude (1>2>3), or alphabetical to denote spatial orientation (v indicates vertical; x and y, horizontal directions respectively perpendicular to and parallel to the BR-CP boundary; capital alphabetical characters, compass directions). Also, it will be useful to define the parameter ϕ as the ratio $(S_2-S_3)/(S_1-S_3)$, expressing a fundamental relation between the relative magnitudes of the principal stresses (see Angelier, 1979); ϕ is defined between limits of 0 and 1.0.

Our inability to select true fault planes and slip directions confidently for the majority of fault-plane solutions, as discussed earlier, causes fundamental problems for applying inversion techniques to solve for stress orientation from the fault-plane solutions (e.g., Angelier, 1984). Assuming, however, that one principal stress is vertical, we can use the data of Figure 20 to argue that the predominance of near-horizontal T-axes for our sample of fault-plane solutions within the general transition zone reflects the orientation of S_3 in the horizontal plane. The mean orientation (±21°) of 30 T-axes having a plunge less than 30° is in the 102°–282° direction. The median orientation is in the 103°–283° direction.

The mixture of normal-faulting and strike-slip focal mechanisms within the transition suggests that S_v, corresponding to the lithostat, interchanges between S_1 and S_2. Accordingly, when P-axes in Figure 20 cluster near the horizontal, they would be interpreted to reflect S_1 in a north-northeast–south-southwest horizontal direction, perpendicular to S_3 in the 102°–282° direction. Otherwise, for S_1 vertical and S_3 horizontal, the P-axes could assume a range of skew orientations in space depending upon ϕ (e.g., Angelier, 1979, Fig. 4). This is because the slip direction (the direction of maximum resolved shear) on a pre-existing fault—and correspondingly the orientations of the auxiliary plane and P-axis—will be a function not only of the orientation of the principal stresses, but also their relative magnitudes.

The average orientation of S_3 that we infer for the BR-CP transition in central and southwest Utah (102°–282°) differs by 25°–30° in a clockwise direction from its average orientation in the Wasatch Front region to the north (075°–255°) (Zoback, 1983), but it is in agreement with the west-northwest regional orientation of S_3 throughout most of the northern BR province (Zoback and Zoback, 1980). Zoback (1983) considered that the east-northeast trend of S_3 in the Wasatch Front region might be influenced by structural grain inherited from a late Precambrian rifted margin. This, coupled with mixed evidence that S_3 assumes a northeasterly orientation in Arizona along the southwestern margin of the CP (Eberhart-Phillips and others, 1981) make it plausible that local orthogonality of S_3 to the BR province boundary is related to deep structure along the boundary (see also Zoback and Zoback, 1980).

McGarr (1982) has calculated changes in stress that should occur in the upper crust across the BR-CP transition—assuming basal tractions resulting from thermally-induced mass transport below the elastic-brittle layer. Our observations are in qualitative agreement with his model predictions, but we recognize that the model is not necessarily unique. First, we find horizontal principal stresses approximately perpendicular and parallel to the province boundary. Second, changes in type of seismic slip across the transition agree with calculated changes in S_x and S_y relative to S_v (i.e., $\rho g z$). Third, an implied high value of ϕ along the BR boundary and the western part of the transition agrees with our observation of azimuthally-clustered T-axes. A large value of ϕ corresponds to S_2 approximately equal to S_1 but much greater than S_3. If ϕ had a very low value in the transition region, as

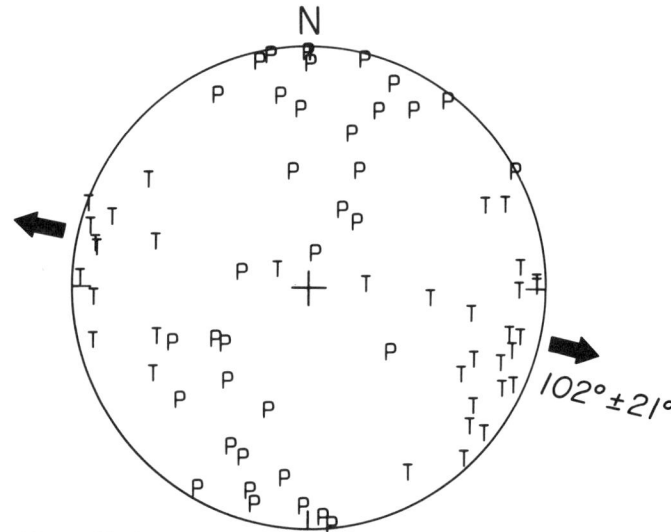

Figure 20. Stereographic plot (lower-hemisphere, equal-area) of P- and T- axes (the standard pressure and tension axes, respectively) for 36 fault-plane solutions shown both in Figure 12 (solutions a–c, h–u) and Figure 15 and lying within the BR-CP transition zone (solutions 1–4, 13, and 26–28 excluded). Large arrows indicate mean orientation of 30 T-axes having a plunge less than 30°; median = 103°.

suggested by Zoback (1984) for the Wasatch Front region, then S_2 should be close in value to S_3 and there should be no preferred orientation for S_3 or the T-axes of the fault-plane solutions. Nor would the mixing of normal and strike-slip faulting be expected.

The occurrence of reverse faulting in the northeastern corner of the transition zone (solutions 24 and 25, Fig. 19) may be explained by locally allowing $S_{NNE}(S_y) > S_{WNW}(S_x) > S_v$. Similarly, solutions 26, 27, and 28 (Fig. 19) could also result from local interchange in values of the principal stresses—in which case the normal faulting associated with solution 28 might not simply reflect eastward continuation of basin-range extension (Humphrey and Wong, 1983).

We add several comments regarding the stress state in the study area. (1) Doser and Smith (1982) computed a moment tensor with a maximum extensional component in the 085°–265° direction based on four moderate-sized earthquakes in southwestern Utah. That tensor orientation, however, depends upon Smith and Sbar's (1974) fault-plane solution for the 1967 Marysvale earthquake, which Doser and Smith (1982) appear to have adopted for three of the four shocks. (2) Anderson and Barnhard (1984) have used extensive Pliocene-Pleistocene fault-slip data from the Sevier Valley area to invert for paleostress and they found the following: S_3 consistently oriented east-west; and S_1 oriented north-south for strike-slip faults and vertical for dipslip faults. (3) From deep hydraulic fracturing in the Chevron #1 Chriss Canyon Well near the southern Wasatch fault (Fig. 7), Arabasz (1984) has deduced a value of S_3 of approximately 750 bars at 5 km depth, corresponding to 0.6 S_v. Assuming hydrostatic pore pressure, in-situ stress differences at the hydrofrac depth are close to the frictional strength of optimally oriented normal faults. (4) A heat flow boundary traced in Fig. 19 separates a "thermal interior" of the Colorado Plateau, characterized by an average heat flow of about 60 mW m^{-2}, from a periphery with average heat flow of about 80–90 mW m^{-2} (Bodell and Chapman, 1982). Whether a distinct stress province within the interior of the Colorado Plateau (Zoback and Zoback, 1980) is also delimited by this boundary remains to be verified by additional stress data, especially in southwestern Utah.

SUMMARY AND CONCLUSIONS

A primary objective of this paper was to summarize detailed seismological observations relating to the BR-CP transition in Utah. The historical absence of large surface-faulting earthquakes and the predominance of diffuse background seismicity represent persistent obstacles for understanding fault behavior and earthquake generation in this region—and, indeed, throughout many other parts of the Intermountain seismic belt. We have emphasized a careful documentation of results from six specific field studies in central and southwestern Utah in which temporary arrays of portable seismographs were used to supplement a regional seismic network. The resulting earthquake hypocenters and fault-plane solutions provide, we believe, the best observational data yet available for deciphering fine details of seismicity in the transition region.

In the preceding three sections we have synthesized new data with other available information to discuss implications for (1) the subsurface geometry of seismically active faults, (2) the correlation of diffuse background seismicity with geological structure, and (3) the nature of a transitional stress state between the BR and CP stress provinces. In terms of relevance to general models of extensional tectonics, we cautioned at the outset that our observations relate, first, to a domain of deformation that is transitional between basin-range extension and plateau uplift, and second, to earthquakes of small-to-moderate size (magnitude <5).

Listric and low-angle normal faults (dips <30°) appear to be widespread in the subsurface along the eastern BR province, and low-angle detachments extend eastward into the transition. Subsurface structure reflects the complicated superposition of Neogene and younger extension (and locally strike-slip faulting) upon pre-Neogene thrustbelt structure. We found no instance in which clustered earthquake foci and corroborating focal mechanisms demonstrate seismic slip on a continuous normal fault that is either downward-flattening or planar with low-angle dip. In the case of a few fault-plane solutions with a low-angle nodal plane we found no independent evidence, such as aftershock clustering on a low-angle plane, that the low-angle plane was indeed the slip plane and not the auxiliary nodal plane. Whether normal slip can occur seismically on a low-angle plane remains to be verified by future experience.

Our studies throughout the BR-CP transition confirm the predominance of seismic slip on fault segments of moderate (>30°) to high-angle dip—at least for small to moderate-sized earthquakes—based on both fault-plane solutions and hypocentral distributions. Mean and median dips of seismic-slip planes for normal-slip fault-plane solutions, both in this study and for northern Utah (Zoback, 1983), range from 49° to 57°. Given independent evidence for the widespread presence of listric and low-angle normal faults in the study area, seismic slip on steeper fault segments might be explained by its being limited to the upper portion of downward-flattening faults, related antithetic faults, or secondary block-interior faults. Such slip on steeper faults may represent adjustment to displacements, either aseismic or seismic, on related low-angle structures. In two areas where we observed background seismicity downdip from the Wasatch fault, earthquake foci and corresponding focal mechanisms are incompatible with seismic slip on either a listric or simple planar projection of the Wasatch fault.

Results of our portable-array earthquake studies lead us to a working hypothesis to explain diffuse background seismicity that may apply not only to the study area but to other parts of the Intermountain seismic belt where crustal structure involves vertically stacked plates separated by low-angle detachments. We suggest that background seismicity is fundamentally controlled by variable mechanical behavior and internal structure of individual horizontal plates within the seismogenic upper crust, as illustrated in Figure 18. Key observations come from the vicinity of the southern Wasatch fault (Fig. 8), the Sevier Valley area (Fig. 11), and the Soda Springs, Idaho area (Fig. 17) where in all cases diffuse background seismicity was observed to predominate above 6–8 km depth. Background earthquake activity at greater depth (down to about 15 km) may be masked or blurred in epicentral view by greater numbers of small- to moderate-sized earthquakes within an overlying plate. Moderate-sized earthquakes up to magnitude 6½ may occur without surface rupture on "blind" subsurface structures that have no direct surface expression. If large surface-faulting earthquakes (magnitude 6½ to 7¾) nucleate at about 15 km depth, as observed elsewhere in the Intermountain region, then rupture pathways must be identified between deep nucleation points and existing surface fault scarps—or the paradox of engaging low-angle detachments seismically must be resolved. In either case, effective seismic surveillance requires precise resolution of focal depths to discriminate seismicity that may be preparatory to large-scale slip.

Fifty-three fault-plane solutions provide significant detail for mapping upper-crustal stress orientation across the BR-CP transition. Important observations, summarized in Figures 19 and 20, include: (1) average regional orientation of S_3, the minimum principal stress, in the 102°–282° direction within the transition; (2) the orientation of horizontal principal stresses perpendicular and parallel to the BR-CP province boundary; and (3) mixed-mode faulting within the transition, including normal slip, strike slip, and oblique slip—and reverse faulting in the northeastern corner of the transition. Intermediate principal stress along the western part of the transition must be closer in value to the maximum principal stress, and these two principal stresses interchange in orientation between vertical and north-northeast–south-southwest.

Finally, perhaps one of the most significant contributions of our portable-array studies throughout the BR-CP transition was the documentation of abundant strike-slip fault-plane solutions in the Sevier Valley area. The experience of making observations in conflict with deeply ingrained assumptions—and having the observations lead to new geological discoveries of strike-slip faulting and a new perspective on an area of classical geology (Anderson and Barnhard, 1984)—was truly rewarding. The BR-CP transition is clearly complex in its history and structure, and our understanding of its seismotectonics is still fragmentary.

ACKNOWLEDGMENTS

This paper synthesizes results from a multi-year program of studies coordinated by the first author. We acknowledge, in particular, significant contributions made by M. E. (McKee) Klingler during an early stage of the program. Also, we gratefully acknowledge joint research efforts with E. McPherson, T. L. Olson, J. C. Pechmann, G. E. Randall, W. D. Richins, and G. Zandt, who were variously involved in the acquisition and/or analysis of data that form the basis of this paper. We thank R. W. Allmendinger, R. E. Anderson, R. L. Bruhn, D. S. Chapman, H. J. Patton, W. G. Powell, F. Royse, R. B. Smith, and L. A. Standlee for helpful discussions; and J. Onstott, D. M. Thomas, and D. J. Williams for help with the manuscript preparation. Critical reviews by D. P. Hill, G. R. Keller, and R. E. Anderson helped improve the manuscript. The multi-year research was supported by: the National Science Foundation, NSF Grant EAR-8008799 (and earlier, EAR-7723706); the U.S. Department of the Interior, Geological Survey, Contract No. 14-08-001-21856 (and predecessor contracts); and the State of Utah.

APPENDIX, EXPERIMENTAL PROCEDURES

Basic data in this paper were obtained from earthquake field studies carried out in the study area between 1979 and 1982 by the University of Utah. The following is a condensed summary of field experimental procedures, with emphasis on the procedure used for determining fault-plane solutions of local earthquakes by the inversion of observed ratios of the amplitudes of SV to P waves. As noted in the text, reliance is placed on other presentations (McKee and Arabasz, 1982; McKee, 1982; Julander, 1983) for more complete details of the seismological data.

Instrumentation

The field studies involved the deployment of smoked-paper-type portable seismographs (up to 10 Sprengnether Model MEQ-800 and two Model MEQ-600 instruments) to supplement the University of Utah's regional seismic telemetry network. Vertical-component, 1-second-period seismometers (Geotech Model S-13 or Kinemetrics Ranger seismometers) were used throughout. During and after 1981, two digital-event-recording seismographs fabricated at the University of Utah were also deployed, each in a three-component mode with Geotech S-13 seismometers. Maximum displacement magnification for the smoked-paper seismographs at 10 Hz ranged from about 1.3×10^5 to 2.0×10^6. Internal crystal clocks synchronized with WWV radio signals provided an accurate time reference. Analog recording was typically at a speed of 60 mm/min allowing timing of impulsive phase arrivals to a precision of ±0.05 sec with routine use of a microscope (Sevier Valley study) or ±0.10 sec with an 8x magnifying ocular.

Field Experiments

Six specific field experiments were carried out in the study area during the following periods: (1) June 3–July 25, 1979 (broad study of easternmost Basin and Range province and western Wasatch Plateau between lat. 39°N and 40°N; 12 portable instruments, 26 sites); (2) July 27–August 14, 1979 (target study of coal-mining-related seismicity in eastern Wasatch Plateau; 10 portable instruments, 13 sites); (3) May 24–28, 1980 (aftershock study following magnitude 4.4 mainshock in Goshen Valley on May 24, 1980; 4 portable instruments, 4 sites—in addition to nearby permanent stations); (4) June 6–July 23, 1981 (systematic study of Sevier Valley between lat. 38°N and 39°N; 13 portable seismographs, 34 sites); (5) May 24–June 1, 1982 (aftershock study following magnitude 4.0 mainshock near Richfield on May 24, 1982; 10 portable seismographs, 10 sites); and (6) June 3–July 28, 1982 (target study of southern Wasatch fault zone; 12 portable seismographs, 16 sites).

During the 1979 and 1980 studies, station spacing for the temporary array was 10 km at best and generally 10–20 km on average. During the subsequent experiments in 1981 and 1982, we adopted a strategy of establishing broad initial coverage (10–15 km station spacing) followed by decreasing station spacing to the order of 5 km to focus on localized zones of earthquake activity.

Hypocenter and Magnitude Determinations

The common procedure for determining earthquake hypocenters was with the computer program HYPOELLIPSE (Lahr, 1979), using both P-wave and S-wave arrival times. P-wave to S-wave velocity ratios (1.65 to 1.76) were determined empirically for each local study area by a linear regression of a large number of pair-wise differences between P-wave and S-wave arrival times. Where indicated in the text, a joint-hypocenter-determination (JHD) technique, developed and described by Herrmann and others (1981), was subsequently applied to refine the HYPOELLIPSE solutions for earthquakes well-recorded with good geometrical control. The JHD technique usefully maximized the spatial precision, i.e., the *relative* location of events (see, for example, Dewey, 1979). A major advantage of the technique is that station adjustments are calculated simultaneously with the hypocentral parameters to minimize the effect of biased travel-time errors that might be due to an inadequate velocity model, poor array geometry, or systematic station anomalies.

Elevation corrections to datum were applied in all cases using either the standard procedure of HYPOELLIPSE or the cited JHD program. (Specific datum values are indicated in the figure captions for all cross sections showing earthquake foci.) Horizontally-layered velocity models appropriate to each local study area were either adapted from published seismic refraction profiles or specifically determined. As part of the Sevier Valley field experiments, extensive efforts were made to refine upper-crustal structure to improve hypocentral accuracy (Julander, 1983). This included use of local quarry blasts as refraction sources and analysis of local earthquake data for multilayering using both the minimum-apparent-velocity method of Matumoto and others (1977) and a simultaneous velocity-hypocenter inversion technique formulated by Benz (1982).

Estimates of local magnitude (M_L) based on signal duration were made for all earthquakes large enough to be recorded by one or more stations of the University of Utah's regional seismic telemetry network (see Arabasz and others, 1979, p. 82 regarding magnitude calibration). Because duration-magnitude scales cannot be extrapolated below about $M_L = 1.5$ without extraordinary calibration (e.g., Suteau and Whitcomb, 1979), events smaller than magnitude 1.5 are simply indicated as such in this paper.

Fault-Plane Solutions

Two basic techniques were used to determine fault-plane solutions in the study. The first, the standard stereographic projection of P-wave first motions is well-known. Except for Solutions 3 and 13 (Figs. 6 and 13), based on regional seismic recordings, only first motions for direct or upward-refracted arrivals were used—to eliminate uncertainties in take-off angle when headwaves are predicted by an assumed velocity model. The second technique, which involves the inversion of observed ratios of the amplitudes of SV to P waves, warrants some description.

Kisslinger (1980) and Feng (1974) have described the theoretical basis for determining the focal mechanism, or fault-plane solution, of a local earthquake from the distribution of ratios of the amplitudes of SV to P, as recorded on vertical-component seismographs. Further, Kisslinger and others (1981) have developed and described an algorithm for computer estimation of focal mechanisms with such data. (Errata to Kisslinger, 1980, and Kisslinger and others, 1981, appear in the *Bulletin of the Seismological Society of America*, v. 72, no. 1., p. 344, February 1982.) Their algorithm, herein referred to as AMPRAT (for amplitude ratio) was extensively used in this study.

The analytical procedures described by Kisslinger and others (1981) were followed. A quantity $(SV/P)_z$ is specified as the ratio of the maximum amplitude of the vertical component of SV to the maximum amplitude of the vertical component of P, where "SV and P amplitudes are read as the maximum peak-to-peak amplitude within the first three half-cycles of the direct arrivals as recorded on the vertical component" (Kisslinger and others, 1981, p. 1721). Values of $(SV/P)_z$ corresponding to ray paths with angles of incidence at the free surface within the range 30° to 37° are rejected because of difficulties in predicting free-surface effects.

The computational procedure of AMPRAT involves specifying a trial value of the slip direction (in terms of angular measurement from the horizontal), if the slip direction is independently known. Alternatively, incremental values of slip between 0° and 180° can be used as starting values. For each starting value of slip direction, AMPRAT then searches an array of strike and dip values to satisfy by iterative least-squares adjustment the observed values of $(SV/P)_z$. Goodness of fit is specified by the sum of squared errors of log $(SV/P)_z$.

To preclude bias regarding an assumed slip direction, values of trial slip between 0° and 180° at 10° increments were used to derive a set of solutions with AMPRAT to best match the observed and predicted values of $(SV/P)_z$ for each selected earthquake.

Criteria for accepting a final solution using AMPRAT included:

(1) Smooth convergence in 9 or preferably fewer iterations (see Kisslinger and others, 1981, regarding the nonlinearity of $(SV/P)_z$ as a function of the parameters being estimated). Confidence in a solution was increased when several initial trial slips converged independently to a consistent estimate.

(2) A root-mean-square *(rms)* error of log $(SV/P)_z$ less than 0.3, which implies matching amplitude ratios within a factor of two.

(3) A well-distributed pattern of $(SV/P)_z$ observations (minimum of 5) with three or more clear first motions that discriminate the compressional and dilational quadrants.

Having searched the solution space with incremental trial values of slip, the solution was selected that had the smallest *rms* error and that satisfied the pattern of P-wave first motions. In some cases, a low *rms*-error solution was accepted with one or two inconsistent first-motion data points if the data points were within 15° of a nodal plane. Summary data for the focal mechanisms determined with AMPRAT are presented in Table 1.

REFERENCES CITED

Aki, K., 1984, Asperities, barriers, characteristic earthquakes and strong motion prediction: Journal of Geophysical Research, v. 89, p. 5867–5872.

Aki, K., and Richards, P. G., 1980, Quantitative seismology—Theory and methods, (2 volumes): San Francisco, W. H. Freeman and Company, 939 p.

Allmendinger, R. W., Sharp, J. W., Von Tish, D., Serpa, L., Brown, L., Kaufman, S., Oliver, J., and Smith, R. B., 1983, Cenozoic and Mesozoic structure of the eastern Basin and Range province, Utah, from COCORP seismic-reflection data: Geology, v. 11, p. 532–536.

Anderson, J. J., and Rowley, P. D., 1975, Cenozoic stratigraphy of southwestern High Plateaus of Utah, *in* Anderson, J. J., Rowley, P. D., Fleck, R. J., and Nairns, A.E.M., eds., Cenozoic geology of southwestern High Plateaus of Utah: Geological Society of America Special Paper 160, p. 1–52.

Anderson, L. W., and Miller, D. G., 1980, Quaternary faulting in Utah, *in* Proceedings, Conference X, Earthquake hazards along the Wasatch and Sierra-Nevada frontal fault zones: U.S. Geological Survey Open-File Report 80-801, p. 194–226.

Anderson, R. E., and Barnhard, T. P., 1984, Extensional and compressional paleostresses and their relationship to paleoseismicity and seismicity, central Sevier Valley, Utah: U.S. Geological Survey Open-File Report 84-763, p. 515–546.

Anderson, R. E., and Buckman, R. C., 1979, Map of fault scarps in unconsolidated sediments, Richfield 1° × 2° Quadrangle, Utah: U.S. Geological Survey Open-File Report 79-1236, 15 p., 1 pl.

Anderson, R. E., Zoback, M. L., and Thompson, G. A., 1983, Implications of selected subsurface data on the structural form and evolution of some basins in the northern Basin and Range Province, Nevada and Utah: Geological Society of America Bulletin, v. 94, p. 1055–1072.

Angelier, J., 1979, Determination of the mean principal directions of stresses for a given fault population: Tectonophysics, v. 56, p. T17–T26.

——, 1984, Tectonic analysis of fault slip data sets: Journal of Geophysical Research, v. 89, p. 5835–5848.

Arabasz, W. J., 1981, Seismicity and listric faulting in central and SW Utah: EOS (American Geophysical Union Transactions), v. 62, p. 960.

——, 1983, Geometry of active faults and seismic deformation within the Basin and Range–Colorado Plateau transition, central and SW Utah: Earthquake Notes, v. 54, no. 1, p. 48.

——, 1984, Swarm seismicity and deep hydraulic fracturing within 10 km of the southern Wasatch fault: Earthquake Notes, v. 55, no. 1, p. 30–31.

Arabasz, W. J., and Robinson, R., 1976, Microseismicity and geologic structure in the northern South Island, New Zealand: New Zealand Journal of Geology and Geophysics, v. 19, p. 569–601.

Arabasz, W. J., and Smith, R. B., 1981, Earthquake prediction in the Intermountain seismic belt—An intraplate extensional regime, *in* Simpson, D. W., and Richards, P. G., eds., Earthquake prediction—An international review: American Geophysical Union, Maurice Ewing Series, v. 4, p. 238–258.

Arabasz, W. J., Smith, R. B., and Richins, W. D., eds., 1979, Earthquake studies in Utah, 1850 to 1978: Salt Lake City, University of Utah Seismograph Stations Special Publication, 552 p.

——, 1980, Earthquake studies along the Wasatch Front, Utah: Network monitoring, seismicity and seismic hazards: Seismological Society of America

Bulletin, v. 70, p. 1479–1499.

Arabasz, W. J., Richins, W. D., and Langer, C. J., 1981, The Pocatello Valley (Idaho-Utah border) earthquake sequence of March to April 1975: Seismological Society America Bulletin, v. 71, p. 803–826.

Armstrong, R. L., 1968, Sevier orogenic belt in Nevada and Utah: Geological Society of America Bulletin, v. 79, p. 429–458.

Benz, H. M., 1982, Simultaneous inversion for lateral velocity variations and hypocenters in the Yellowstone region using earthquake and refraction data [M.S. thesis]: Salt Lake City, University of Utah, 105 p.

Best, M. G., and Hamblin, W. K., 1978, Origin of the northern Basin and Range province: Implications from the geology of its eastern boundary, in Smith, R. B., and Eaton, G., eds., Cenozoic tectonics and regional geophysics of the western Cordillera: Geological Society of America Memoir 152, p. 313–340.

Bodell, J. M., and Chapman, D. S., 1982, Heat flow in the north-central Colorado Plateau: Journal of Geophysical Research, v. 87, p. 2869–2884.

Brown, R. P., and Cook, K. L., 1982, A regional gravity survey of the Sanpete-Sevier valleys and adjacent areas in Utah, in Nielson, D. L., ed., Overthrust belt of Utah: Salt Lake City, Utah Geological Association Publication 10, p. 121–135.

Brun, J.-P., Choukroune, P., 1983, Normal faulting, block tilting, and decollement in a stretched crust: Tectonics, v. 2, p. 345–356.

Bucknam, R. C., Algermissen, S. T., and Anderson, R. E., 1980, Patterns of late Quaternary faulting in western Utah and application in earthquake hazard evaluation, in Proceedings, Conference X, Earthquake hazards along the Wasatch and Sierra-Nevada frontal fault zones: U.S. Geological Survey Open-File Report 80-801, p. 299–314.

Burchfiel, B. C., and Hickcox, C. W., 1972, Structural development of central Utah, in Baer, J. L., and Callaghan, E., eds., Plateau–Basin and Range transition zone, central Utah, 1972: Salt Lake City, Utah Geological Association Publication 2, p. 55–64.

Chapin, C. E., and Cather, S. M., 1981, Eocene tectonics and sedimentation in the Colorado Plateau–Rocky Mountain area, in Dickinson, W. R., and Payne, W. D., eds., Relations of tectonics to ore deposits in the southern Cordillera: Arizona Geological Society Digest, v. 14, p. 173–198.

Cook, K. L., and Smith, R. B., 1967, Seismicity in Utah, 1850 through June 1965: Seismological Society of America Bulletin, v. 57, p. 689–718.

Crone, A. J., and Harding, S. T., 1984, Relationship of late Quaternary fault scarps to subjacent faults, eastern Great Basin, Utah: Geology, v. 12, p. 292–295.

Das, S., and Scholz, C. H., 1981, Off-fault aftershock clusters caused by shear stress increase?: Seismological Society of America Bulletin, v. 71, p. 1669–1675.

—— , 1983, Why large earthquakes do not nucleate at shallow depths: Nature, v. 305, p. 621–623.

Davis, D. A., 1983, Gravity survey of Utah and Goshen Valley and adjacent areas, Utah [M.S. thesis]: Salt Lake City, University of Utah, 141 p.

Davis, G. H., 1978, Monocline fold pattern of the Colorado Plateau, in Matthews, V., III, ed., Laramide folding associated with basement block faulting in the western United States: Geological Society of America Memoir 151, p. 215–234.

Dewey, J. W., 1979, A consumer's guide to instrumental methods for determination of hypocenters: Geological Society of America Reviews in Engineering Geology, v. 4, p. 109–117.

Dixon, J. S., 1982, Regional structural synthesis, Wyoming salient of western overthrust belt: American Association of Petroleum Geologists Bulletin, v. 66, p. 1560–1580.

Doelling, H. H., 1972, Central Utah coal fields: Sevier-Sanpete, Wasatch Plateau, Book Cliffs, and Emery, Utah: Salt Lake City, Utah Geological and Mineralogical Survey Monograph 3, 571 p.

Doser, D. I., 1985, The 1983 Borah Peak, Idaho and 1959 Hebgen Lake, Montana earthquakes: Models for normal fault earthquakes in the Intermountain seismic belt: U.S. Geological Survey Open-File Report 85-290, p. 368–384.

Doser, D. I., and Smith, R. B., 1982, Seismic moment rates in the Utah region: Seismological Society of America Bulletin, v. 72, p. 525–551.

Eberhart-Phillips, D., Richardson, R. M., Sbar, M. L., and Herrmann, R. B., 1981, Analysis of the 4 February 1976 Chino Valley, Arizona, earthquake: Seismological Society of America Bulletin, v. 71, p. 787–801.

Feng, T. Y., 1974, Anomalies of amplitude ratio of S and P waves from near earthquakes and earthquake prediction: Acta Geophysica Sinica, v. 17, p. 140–153 (translated into English in Chineses Geophysics, 1978, v. 1, p. 1–15).

Fenneman, N. M., 1928, Physiographic divisions of the United States: Annals of the Association of American Geographers, v. 18, p. 261–353.

Gibbs, A. D., 1983, Balanced cross-section construction from seismic sections in areas of extensional tectonics: Journal of Structural Geology, v. 5, p. 153–160.

Gibowicz, S. J., Droste, Z., Guterch, B., and Hordejuk, J., 1981, The Belchatow, Poland, earthquakes of 1979 and 1980 induced by surface mining: Engineering Geology, v. 17, p. 257–271.

Gries, R., 1983, North-south compression of Rocky Mountain foreland structures, in Lowell, J. D., and Gries, R., eds., Rocky Mountain foreland basins and uplifts: Denver, Rocky Mountain Association of Geologists, v. p. 9–32.

Halliday, M. E., and Cook, K. L., 1980, Regional gravity survey, northern Marysvale volcanic field, south-central Utah: Geological Society of America Bulletin, v. 91, p. 502–508.

Hamblin, W. K., 1984, Direction of absolute movement along the boundary faults of the Basin and Range–Colorado Plateau margin: Geology, v. 12, p. 116–119.

Herrmann, R. B., Park, S.-K., and Wang C.-Y., 1981, The Denver earthquakes of 1967–1968: Seismological Society of America Bulletin, v. 71, p. 731–745.

Hill, D. P., 1982, Contemporary block tectonics: California and Nevada: Journal of Geophysical Research, v. 87, p. 5433–5450.

Hintze, L. F., 1980, Geologic map of Utah: Salt Lake City, Utah Geological and Mineral Survey Map, scale 1:500,000.

Hulen, J. B., and Sandberg, M., 1981, Exploration case history of the Monroe KGRA, Sevier County, Utah: Salt Lake City, University of Utah Research Institute, Earth Science Laboratory Technical Report DOE/ID/12079-11, ESL 149, 82 p.

Humphrey, J. R., and Wong, I. G., 1983, Recent seismicity near Capitol Reef National Park, Utah, and its tectonic implications: Geology, v. 11, p. 447–451.

Hunt, C. B., 1956, Cenozoic geology of the Colorado Plateau: U.S. Geological Survey Professonal Paper 279, 95 p.

Jackson, J., and McKenzie, D., 1983, The geometrical evolution of normal fault systems: Journal of Structural Geology, v. 5, p. 471–482.

Jaeger, J. C., and Cook, N.G.W., 1979, Fundamentals of rock mechanics (third edition): London, Chapman and Hall, 593 p.

Julander, D. R., 1983, Seismicity and correlation with fine structure in the Sevier Valley area of the Basin and Range–Colorado Plateau transition, south-central Utah (M.S. thesis): Salt Lake City, University of Utah, 143 p.

Kanamori, H., 1981, The nature of seismicity patterns before large earthquakes, in Simpson, D. W., and Richards, P. G., eds., Earthquake prediciton—An international review: American Geophysical Union, Maurice Ewing Series, v. 4, p. 1–19.

Kastrinsky, A. J., 1977, Seismicity of the Wasatch Front, Utah: Detailed epicentral patterns and anomalous activity (M.S. thesis): Salt Lake City, University of Utah, 139 p.

Keller, G. R., Braile, L. W., and Morgan, P., 1979, Crustal structure, geophysical models and contemporary tectonism of the Colorado Plateau: Tectonophysics, v. 61, p. 131–147.

Kisslinger, C., 1980, Evaluation of S to P amplitude ratios for determining focal mechanisms from regional network observations: Seismological Society of America Bulletin, v. 70, p. 999–1014.

Kisslinger, C., Bowman, J. R., and Koch, K., 1981, Procedures for computing focal mechanisms from local (SV/P)z data: Seismological Society of American Bulletin, v. 71, p. 1719–1729.

Kligfield, R., Crespi, J., Naruk, S., and Davis, G. H., 1984, Displacement and strain patterns of extensional orogens: Tectonics, v. 3, p. 577–609.

Lahr, J. C., 1979, HYPOELLIPSE: A computer program for determining local earthquake hypocentral parameters, magnitude, and first motion pattern: U.S. Geological Survey Open-File Report 79-431, 53 p.

Lay, T., and Kanamori, H., 1981, An asperity model of large earthquake sequences, in Simpson, D. W., and Richards, P. G., eds., Earthquake prediction—An international review: American Geophysical Union, Maurice Ewing Series, v. 4, p. 579–592.

Leith, W., Simpson, D. W., and Alvarez, W., 1981, Structure and permeability: Geologic controls on induced seismicity at Nurek reservoir, Tadjikistan, USSR: Geology, v. 9, p. 440–444.

Machette, M. N., 1982, Beaver Basin, south-central Utah, in Scott, W. E., Machette, M. N., Shroba, R. R., and McCoy, W. D., eds., Guidebook for the 1982 Friends of the Pleistocene Rocky Mountain Cell Field Trip to Central Utah, Part II, p. 1–42 (see also U.S. Geological Survey Open-File Report 82-850).

Matumoto, T., Ohtake, M., Latham, G., and Umana, J., 1977, Crustal structure in southern Central America: Seismological Society of America Bulletin, v. 67, p. 121–134.

McDonald, R. E., 1976, Tertiary tectonics and sedimentary rocks along the transition, Basin and Range province to Plateau and thrust belt province, Utah, in Hill, J. G., ed., Symposium on Geology of the Cordilleran hingeline: Denver, Rocky Mountain Association of Geologists, p. 281–317.

McGarr, A., 1982, Analysis of states of stress between provinces of constant stress: Journal of Geophysical Research, v. 87, p. 9279–9288.

McKee, M. E., 1982, Microearthquake studies across the Basin and Range–Colorado Plateau transition zone in central Utah (M.S. thesis): Salt Lake City, University of Utah, 118 p.

McKee, M. E., and Arabasz, W. J., 1982, Microearthquake studies across the Basin and Range–Colorado Plateau transition in central Utah, in Nielson, D. L., ed., Overthrust belt of Utah: Salt Lake City, Utah Geological Association Publication 10, p. 137–149.

Olson, T. L., 1976, Earthquakes surveys of the Roosevelt Hot Springs and the Cove Fort area, Utah [M.S. thesis]: Salt Lake City, University of Utah, 81 p.

Pack, F. J., 1921, The Elsinore earthquakes in central Utah, September 29 and October 1, 1921: Seismological Society of America Bulletin, v. 11, p. 155–165.

Patton, H. J., 1985, P-wave fault-plane solutions and the generations of surface waves by earthquakes in the western United States: Geophysical Research Letters, v. 12, p. 518–521.

Pechmann, J. C., and Thorbjarnardottir, B., 1984, Investigations of an M_L 4.3 earthquake in the western Salt Lake Valley using digital seismic data: U.S. Geological Survey Open-File Report 84-763, p. 340–365.

Powell, W. G., Chapman, D. S., and Bodell, J. M., 1983, Heat flow in the Colorado Plateau of southern Utah: EOS (American Geophysical Union Transactions), v. 64, p. 836.

Richins, W. D., Arabasz, W. J., Hathaway, G. M., Oehmich, P. J., Sells, L. L., and Zandt, G., 1981a, Earthquake data for the Utah Region: July 1, 1978 to December 31, 1980: Salt Lake City, University of Utah Seismograph Stations Special Publication, 125 p.

Richins, W. D., Zandt, G., and Arabasz, W. J., 1981b, Swarm seismicity along the Hurricane fault zone during 1980–81: A typical example for SW Utah: EOS (American Geophysical Union Transactions), v. 62, p. 966.

Richins, W. D., Arabasz, W. J., and Langer, C. J., 1983, Episodic earthquake swarms ($M_L \leq 4.7$) near Soda Springs, Idaho, 1981–82: Correlation with local structure and regional tectonics: Earthquake Notes, v. 54, no. 1, p. 99.

Richins, W. D., Arabasz, W. J., Hathaway, G. M., McPherson, E., Oehmich, P. J., and Sells, L. L., 1984, Earthquake data for the Utah Region: January 1, 1981 to December 31, 1983: Salt Lake City, University of Utah Seismograph Stations Special Publication, 111 p.

Rowley, P. D., Anderson, J. J., Williams, P. L., and Fleck, R. J., 1978, Age of structural differentiation between the Colorado Plateau and Basin and Range provinces in southwestern Utah: Geology, v. 6, p. 51–55.

Rowley, P. D., Steven, T. A., Anderson, J. J., and Cunningham, C. C., 1979, Cenozoic stratigraphic and structural framework of southwestern Utah: U.S. Geological Survey Professional Paper 1149, 22 p.

Rowley, P. D., Steven, T. A., and Mehnert, H. H., 1981, Origin and structural implications of upper Miocene rhyolites in Kingston Canyon, Piute County, Utah: Geological Society of America Bulletin, v. 92, p. 590–602.

Royse, F., 1983, Extensional faults and folds in the foreland thrust belt, Utah, Wyoming, Idaho: Geological Society of America Abstracts with Programs, v. 15, no. 5, p. 295.

Sbar, M. L., 1982, Delineation and interpretation of seismotectonic domains in western North America: Journal of Geophysical Research, v. 87, p. 3919–3928.

Sbar, M. L., Barazangi, M., Dorman, J., Scholz, C. H., and Smith, R. B., 1972, Tectonics of the Intermountain seismic belt, western United States: Microearthquake seismicity and composite fault plane solutions: Geological Society of America Bulletin, v. 83, p. 13–28.

Scholz, C. H., 1968, Experimental study of the fracturing process in brittle rocks: Journal of Geophysical Research, v. 73, p. 1447–1454.

Schwartz, D. P., and Coppersmith, K. J., 1984, Fault behavior and characteristic earthquakes: Example from the Wasatch and San Andreas fault zones: Journal of Geophysical Research, v. 89, p. 5681–5698.

Segall, P., and Pollard, D. D., 1980, Mechanics of discontinuous faults: Journal of Geophysical Research, v. 85, p. 4337–4350.

Shuey, R. T., Schellinger, D. K., Johnson, E. G., and Alley, L. B., 1973, Aeromagnetics and the transition between the Colorado Plateau and Basin and Range provinces: Geology, v. 1, p. 107–110.

Sibson, R. H., 1982, Fault zone models, heat flow, and the depth distribution of earthquakes in the continental crust of the United States: Seismological Society of America Bulletin, v. 72, p. 151–163.

Smith, R. B., 1978, Seismicity, crustal structure, and intraplate tectonics of the interior of the western Cordillera, in Smith, R. B., and Eaton, G., eds., Cenozoic tectonics and regional geophysics of the western Cordillera: Geological Society of America Memoir 152, p. 111–144.

Smith, R. B., and Bruhn, R. L., 1984, Intraplate extensional tectonics of the eastern Basin-Range: Inferences on structural style from seismic reflection data, regional tectonics, and thermal-mechanical models of brittle-ductile deformation: Journal of Geophysical Research, v. 89, p. 5733–5762.

Smith, R. B., and Lindh, A., 1978, A compilation of fault plane solutions of the western United States, in Smith, R. B., and Eaton, G. P., eds., Cenozoic tectonics and regional geophysics of the western Cordillera: Geological Society of America Memoir 152, p. 107–110.

Smith, R. B., and Sbar, M. L., 1974, Contemporary tectonics and seismicity of the western United States with emphasis on the Intermountain seismic belt: Geological Society of America Bulletin, v. 85, p. 1205–1218.

Smith, R. B., Winkler, P. L., Anderson, J. G., and Scholz, C. H., 1974, Source mechanisms of microearthquakes associated with underground mines in eastern Utah: Seismological Society of America Bulletin, v. 64, p. 1295–1317.

Smith, R. B., Richins, W. D., Doser, D. I., Pechmann, J. C., Leu, L. L., and Chen, G. J., 1984, The 1983, M_s 7.3 Borah Peak, Idaho earthquake: A model for active crustal extension: EOS (American Geophysical Union Transactions), v. 65, p. 989.

Spieker, E. M., 1949, The transition between the Colorado Plateau and the Great Basin in central Utah: Salt Lake City, Utah Geological Society, Guidebook to the geology of Utah, no. 4, 106 p.

Standlee, L. A., 1982, Structure and stratigraphy of Jurassic rocks in central Utah: Their influence on tectonic development of the Cordilleran foreland and thrust belt, in Powers, R. B., ed., Geologic studies of the Cordilleran thrust belt, v. 1: Denver, Rocky Mountain Association of Geologists, p. 357–382.

Steven, T. A., Rowley, P. D., and Cunningham, C. G., 1984, Calderas of the Marysvale volcanic field, west central Utah: Journal of Geophysical Research, v. 89, p. 8751–8764.

Stewart, J. H., Moore, W. J., and Zietz, I., 1977, East-west patterns of Cenozoic igneous rocks, aeromagnetic anomalies and mineral deposits, Nevada and Utah: Geological Society of America Bulletin, v. 88, p. 67–77.

Stokes, W. L., 1976, What is the Wasatch Line?, in Hill, J. G., ed., Geology of the

Cordilleran hingeline: Denver, Rocky Mountain Association of Geologists, p. 11–25.
——, 1977, Subdivisions of the major physiographic provinces in Utah: Utah Geology, v. 4, no. 1, p. 1–17.
——, 1979, Stratigraphy of the Great Basin region, *in* Basin and Range symposium and Great Basin field conference, 1979: Denver, Rocky Mountain Association of Geologists–Utah Geological Association Guidebook, p. 195–219.
Suteau, A. M., and Whitcomb, J. H., 1979, A local earthquake coda magnitude and its relation to deviation, moment M_o, and local Richter magnitude M_L: Seismological Society of America Bulletin, v. 69, p. 353–368.
Swan, F. H., III, Schwartz, D. P., and Cluff, L. S., 1980, Recurrence of moderate-to-large magnitude earthquakes produced by surface faulting on the Wasatch fault zone, Utah: Seismological Society of America Bulletin, v. 70, p. 1431–1462.
Thompson, G. A., and Zoback, M. L., 1979, Regional geophysics of the Colorado Plateau: Tectonophysics, v. 61, p. 149–181.
Vetter, U., and Ryall, A. S., 1983, Systematic change of focal mechanism with depth in the western Great Basin: Journal of Geophysical Research, v. 88, p. 8237–8250.
Wechsler, D. J., 1979, An evaluation of hypocenter location techniques with applications to southern Utah: Regional earthquake distributions and seismicity of geothermal areas [M.S. thesis]: Salt Lake City, University of Utah, 225 p.
Wernicke, B., and Burchfiel, B. C., 1982, Modes of extensional tectonics: Journal of Structural Geology, v. 4, p. 105–115.
Westphal, W. H., and Lange, A. L., 1966, Local seismic monitoring: Menlo Park, California, Stanford Research Institute Technical Report, 242 p.
Williams, J. S., and Tapper, M. L., 1953, Earthquake history of Utah, 1850–1949: Seismological Society of America Bulletin, v. 43, p. 191–218.
Witkind, I. J., 1982, Salt diapirism in central Utah, *in* Nielson, D. L., ed., Overthrust belt of Utah: Salt Lake City, Utah Geological Association Publication 10, p. 13–30.
Wong, I. G., 1984, Mining-induced seismicity in the Colorado Plateau, western United States, and its implications for the siting of an underground high-level nuclear waste repository, *in* Gay, N. C., and Wainwright, E. H., eds., Proceedings of the 1st International Congress on Rockbursts and Seismicity in Mines: Johannesburg, South African Institute of Mining and Metallurgy, p. 147–152.
Wong, I. G., and Simon, R. B., 1981, Low-level historical and contemporary seismicity in the Paradox Basin, Utah and its tectonic implications, *in* Wiegand, D., ed., Guidebook to 1981 field conference: Denver, Rocky Mountain Association of Geologists, p. 169–183.
Zandt, G., McPherson, L., Schaff, S., and Olsen, S., 1982, Seismic baseline and induction studies, Roosevelt Hot Springs, Utah and Raft River Idaho: Salt Lake City, University of Utah Research Institute, Earth Science Laboratory Technical Report, Department of Energy Contract No. DE-AS07-78ID01821, 58 p.
Zoback, M. L., 1983, Structure and Cenozoic tectonism along the Wasatch fault zone, Utah, *in* Miller, D. M., Todd, V. R., and Howard, K. A., eds., Tectonics and stratigraphy of the eastern Great Basin: Geological Society of America Memoir 157, p. 3–27.
——, 1984, Constraints on the in-situ stress field along the Wasatch Front: U.S. Geological Survey Open-File Report 84-763, p. 286–309.
Zoback, M. L., and Zoback, M. D., 1980, State of stress in the con-terminous United States: Journal of Geophysical Research, v. 85, p. 6113–6156.
Zoback, M. L., Anderson, R. E., and Thompson, G. A., 1981, Cainozoic evolution of the state of stress and style of tectonism of the Basin and Range province of the western United States: Philosophical Transactions of the Royal Society of London, ser. A, v. 300, p. 407–434.

MANUSCRIPT ACCEPTED BY THE SOCIETY MARCH 4, 1986

Patterns and modes of early Miocene crustal extension, central Mojave Desert, California

Roy K. Dokka
Department of Geology,
Louisiana State University,
Baton Rouge, Louisiana 70803

ABSTRACT

The upper crust of the central Mojave Desert was extended and thinned during early Miocene time by three processes: (1) low-angle normal faulting; (2) high-angle normal faulting in several episodes; and (3) extension fracturing. The first two processes appear to have affected the entire breadth of the extended area, but the third process was restricted to narrow zones. Processes 1 and 2 developed in at least two half-grabens that were kinematically linked by a transform fault. Each half-graben contains a family of similarly oriented high-angle normal faults that facilitated the extension of the upper crust. Evidence from one of the half-grabens suggests that the high-angle faults originally dipped at higher angles and were mechanically linked with a gently northeast-dipping normal fault (detachment).

INTRODUCTION

Great strides have been made during the past two decades in improving our understanding of the nature of continental rifting. This has come to pass primarily because of advances made at two different levels of observation. Acquisition and integration of geologic and geophysical data from ocean basins have given us new insights into the structural development of rifts on a global and regional scale. Although this line of investigation has provided a better understanding of the tectonic context in which continental rifts develop and has identified possible controls on continental rifting, it has not elucidated the details of the specific processes by which continental extension takes place. Many recent advances in this area have been made by field geologists conducting detailed structural studies in the southwestern United States, particularly in the Basin and Range province. Attracted by the possibility of observing deep structural levels of basin-range faults exposed by the Colorado River south of Las Vegas, Nevada, R. E. Anderson (1971) first recognized the importance of low-angle faults as a mode of crustal extension. His pioneering work has caused many geologists and geophysicists to reevaluate their thinking regarding the processes of continental rifting, and has inspired them to go back to the field to gather new data so that a more exact understanding of extensional orogenesis can be obtained.

This paper is a descriptive and kinematic synthesis of one such extended area that developed in the central Mojave Desert of southern California during early Miocene time. This discussion of the central Mojave extensional complex (CMEC) is based primarily on detailed field mapping carried out at scales of 1:12,000 and larger. The first objective of this study was to document the geometry, kinematics, and timing of development of the extension-related structural elements within the CMEC. The second objective of this study was to develop a three-dimensional model to explain the geometric and kinematic relationships observed in the CMEC.

GEOLOGIC SETTING OF THE CENTRAL MOJAVE DESERT

The Mojave Desert block of southern California is a triangle-shaped structural province bounded by the Garlock fault to the north and the San Andreas and Pinto Mountain fault systems to the south. Its eastern limit is marked by the Death Valley and Granite Mountains faults (Fig. 1) (Dokka, 1983a).

The greater Mojave Desert area is composed of the following lithological assemblages: (1) Precambrian complex (metamorphic and igneous rocks) ranging in age from 1.87 Ga to 1.2 Ga (Burchfiel and Davis, 1980); (2) upper Precambrian-

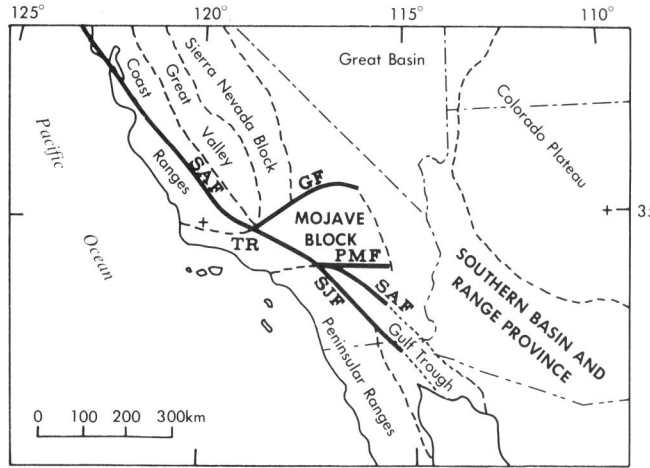

Figure 1. Index map showing location of the Mojave Desert. SAF = San Andreas fault; GF = Garlock fault; PMF = Pinto Mountain fault; SJF = San Jacinto fault; TR = Transverse Ranges.

Paleozoic sequence of miogeoclinal and platform rocks (Stewart and Poole, 1975; Burchfiel and Davis, 1980); (3) Mesozoic back-arc and intra-arc shallow-marine and continental sequences (Burchfiel and Davis, 1980); (4) Mesozoic batholithic rocks and associated (?) volcanic cover (Kistler, 1974; Burchfiel and Davis, 1980); (5) lower Miocene calc-alkaline and bimodal volcanic sequences and continental sedimentary rocks (Dibblee, 1967a; Armstrong and Higgins, 1973; Woodburne and others, 1982); (6) middle Miocene and younger continental sedimentary rocks and intermediate to silicic volcanic rocks (Dibblee, 1967a; Woodburne and others, 1982; Woodburne and Tedford, 1982); and (7) Quaternary alkalic volcanic rocks (Wise, 1969) and unconsolidated sediments.

The Mojave Desert block was affected by at least four tectonic events during Cenozoic time. The first disturbance was a broad uplift of the entire block that occurred at some time prior to the Miocene Epoch. Little is known about this event because of the lack of lower Cenozoic rocks. Hewett (1954) estimated that approximately 4.5 km of erosion may have occurred in association with this uplift as evidenced by the absence of sedimentary rocks that were probably deposited across the entire Mojave region during late Precambrian, Paleozoic, and Mesozoic times (Stewart and Poole, 1975; Tyler, 1979; E. Miller, 1977; Burchfiel and Davis, 1980). At the beginning of Miocene time the central Mojave Desert underwent a profound change in physiography, tectonics, magmatism, and sedimentation patterns as a result of a short, but intense interval of crustal extension (Dokka, 1979, 1980, this volume). Extension along a west-southwest–east-northeast line occurred between 23 and 20 Ma and affected an area in excess of 5 000 km². Extension of the Mojave area coincided with initiation of Tertiary magmatism (Armstrong and Higgins, 1973). Minor extension may have persisted until 16–17 Ma (Dokka, 1979) when the east-west-trending Barstow Basin was established across the western and central Mojave Desert

(Dibblee, 1967a; Burke and others, 1982; Woodburne and others, 1982). The Barstow Basin contains as much as 1000 m of continental sediments, and it persisted until ~13 Ma (Dibblee, 1967a; Burke and others, 1982). The tectonic controls on the formation of this basin are not understood. Right strike-slip faulting, presumably related to Pacific–North American plate interaction, developed in late Tertiary time (Dibblee, 1961; Garfunkel, 1974; Dokka, 1983a).

STRUCTURE OF THE CMEC

The central Mojave Desert was extended during early Miocene time by a combination of extensional processes that included: (1) low-angle normal (detachment) faulting; (2) high-angle normal faulting; and (3) extensional fracturing (with associated dike emplacement and vein growth). All of the structures that are currently exposed formed in the brittle field; there is no evidence for kinematically related ductile deformation as has been documented in some Cordilleran extensional belts (e.g., Rehrig and Reynolds, 1980). This does not rule out the possibility, however, that deeper structural levels in the Mojave region may have been extended concurrently by ductile flow.

Two domains of extension can be recognized within the CMEC on the basis of differences in faulting style, fault geometry, and associated stratal tilt pattern (Fig. 2). The largest of the two domains, here named the Daggett terrane, includes rocks of the Newberry Mountains, southern and central Cady Mountains, northern Bristol Mountains, northern Lava Hills, and the northernmost Bullion Mountains. A second domain, the Bullion terrane, occurs south of the Daggett terrane and includes the central and southern parts of the Bullion Mountains and the Lava Bed Mountains.

Daggett Terrane

The structure of this terrane is dominated by a regional (?) low-angle normal fault (detachment) and a family of uniformly tilted blocks bounded by northwest-striking, moderately northeast-dipping normal faults. Although a complete understanding of the mechanical relationship between the two fault families is hampered by low structural relief in the area, field relations suggest that: (1) the detachment and the high-angle faults were kinematically related; and (2) the upper-plate fault blocks behaved in a domino-like fashion (planar-fault bounded) (e.g., Thompson, 1960). The major phase of extension was preceded by an interval of dike emplacement and was succeeded by a minor interval of high-angle normal faulting.

Newberry Mountains Detachment Fault. The Newberry Mountains detachment fault (NMDF) is a regional low-angle normal fault that separates an extended upper plate from a non-metamophosed and relatively undeformed lower plate. Although considered to be a buttress unconformity of colossal proportions by Dibblee (1970, 1971), stratigraphic relations and fault-plane features leave no doubt as to the dislocational nature of this

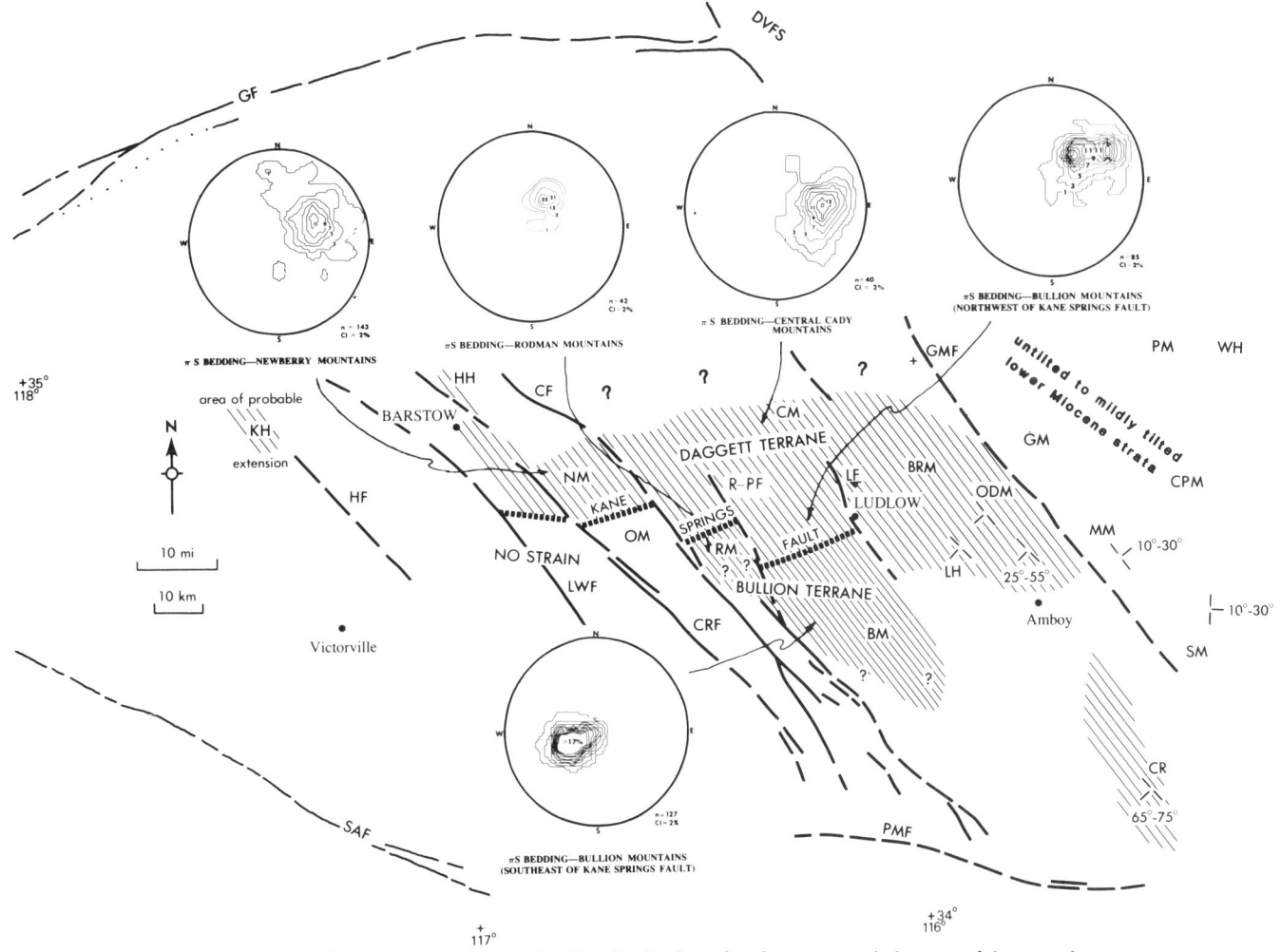

Figure 2. Location and structure map showing distribution of surface structural elements of the central Mojave extensional complex (CMEC). Northwest-striking faults are young right-slip faults (HF = Helendale fault; LWF = Lenwood fault; CRF = Camp Rock fault; CF = Calico fault; R-PF = Rodman-Pisgah fault; LF = Ludlow fault). Stereograms are π diagrams of bedding in lower Miocene strata. Regional faults are abbreviated as follows: GF = Garlock fault; SAF = San Andreas fault; DVFS = Death Valley fault zone; PMF = Pinto Mountain fault. Locations are keyed to text: KH = Kramer Hills; HH = Hinkley Hills; NM = Newberry Mountains; OM = Ord Mountains; RM = Rodman Mountains; CM = Cady Mountains; CR = Calumet Range; GM = Granite Mountains; MM = Marble Mountains; SM = Ship Mountains; CPM = Clipper Mountains; PM = Piute Mountains; WH = Wood Hills; ODM = Old Dad Mountains; BRM = Bristol Mountains; LH = Lava Hills. Stratal orientations from southeastern areas are from Miller and others (1982) and Bassett and Kupfer (1964). A line coincident with the trace of the Granite Mountains fault (GMF) (Dokka, 1983a, 1983b) separates the CMEC from untilted and slightly tilted lower Miocene rocks to the east.

contact. For example, sedimentary and volcanic rocks juxtaposed across this zone do not contain any clasts of adjacent lower-plate rocks (i.e., cataclasized Mesozoic granite-quartz monzonite or Mesozoic volcanic rocks (Sidewinder Volcanic Series), nor do the strata onlap the basement as might be expected if this surface were one of deposition as suggested by Dibblee (1970, 1971). Instead, this contact is marked by a zone of extreme shearing and shattering of lower-plate rocks, comminution of materials near the fault, development of fault-plane features and kinematic indicators (slickensides, striations, and fibrous mineral growths), and the abrupt truncation of bedding and foliations (Fig. 3).

Mapping of this surface strongly suggests that it once continuously underlay the Newberry Mountains. Although the NMDF does not crop out in other ranges in the Daggett terrane, its presence in the subsurface is suspected because of the similarity of upper-plate structure (tilted and extended lower Miocene strata and basement rocks) to that of the Newberry Mountains (Fig. 2). Because of disruption by later faulting and doming, the NMDF now occurs as a series of smaller sheets (Fig. 3). Individual fault segments strike northwest and dip gently (0–20°) to the southwest. Near Su Casa dome, the fault has been upwarped into an east-west–trending, doubly plunging antiform (Fig. 3) that may

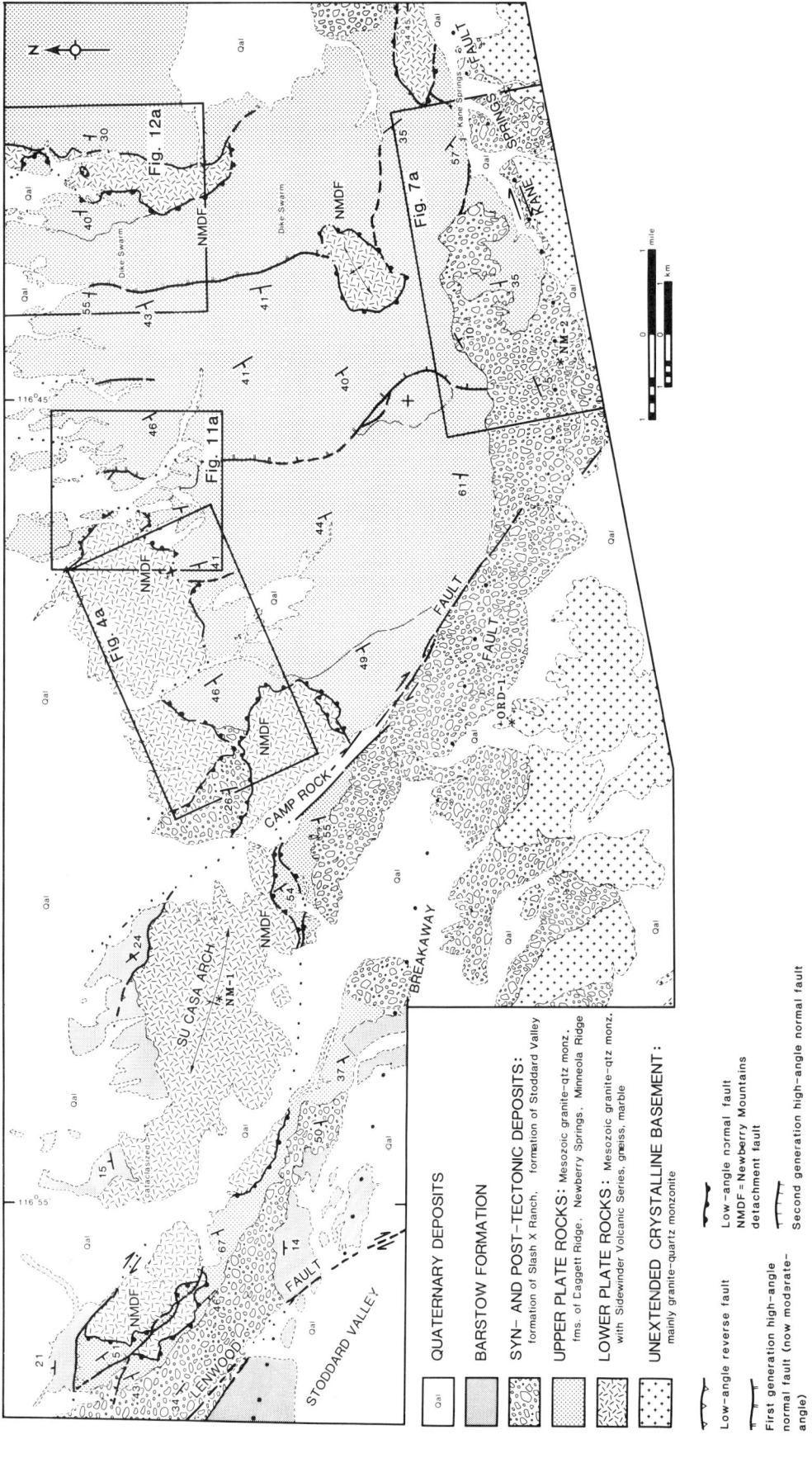

Figure 3. Generalized geologic map of the Newberry Mountains highlighting the locations of the Newberry Mountains detachment fault (NMDF), Kane Springs fault, upper-plate normal faults, and the breakaway fault. Locations of maps given in Figures 4, 7, 11, and 12 are noted. Locality names keyed to text.

be analogous to arches associated with middle Tertiary extensional terranes of the Colorado River trough (Spencer, 1984). The time of doming at Su Casa has been bracketed between 20 and 16 Ma on the basis of stratigraphic relations (Dokka, 1980).

Map relations and cross sections indicate that the NMDF is locally fluted (i.e., downdip profiles of the fault are curviplanar or steplike). The best example of this is found near the Azucar Mine (Fig. 4) where the NMDF changes orientation from subhorizontal to northwest-striking and moderately southwest-dipping. Inasmuch as the nearby strata are not folded, upwarping by later doming cannot alone be invoked to explain the curvature of this part of the fault. The NMDF must have initially formed as a curved surface.

The NMDF crops out discontinuously along its trace and is generally marked by a resistant ledge of cataclastic rock and gouge that caps a variably thick (5–100 m and perhaps more) zone of coherent microbreccia and cataclasite (nomenclature after Higgins, 1971) (Figs. 5 and 6). These rocks are considered to be related to the NMDF because: (1) the zone of cataclasis only occurs adjacent to the NMDF; and (2) the unit contains shear zones that are geometrically and kinematically similar to the NMDF.

The structure of the cataclastic zone is dominated by anastomosing subhorizontal shears and north-northwest–striking, low-angle fractures. Cataclastic rocks can be seen to gradually give way to noncataclastic rock at depth. Figure 5 is a series of photographs through a single exposure of cataclastic granite on the southern flank of the Su Casa dome in the western Newberry Mountains. They illustrate the gradational character of shearing effects through the cataclastic zone.

Figure 5a is located 50 m below the NMDF and shows a rock that has undergone mild cataclasis. The only structures present are a set of low-angle shear zones and a family of moderately dipping faults. Comminution of the rock is volumetrically insignificant and restricted to the shear zones. Zones of cataclasis are separated by slabs of undeformed basement rocks of variable thickness; the thickness of these slabs tends to decrease toward the NMDF. The intermediate stage of cataclasis of these rocks is shown on Figure 5b. The base of this subunit is defined by the disappearance of undeformed slabs of basement like those seen below. At this intermediate stage of cataclasis, comminuted material is volumetrically important, phacoid-shaped porphyroclasts begin to appear, and fault density increases. Near the NMDF (Fig. 5c and 6), porphyroclasts tend to be matrix supported because of the high proportion of pulverized material. The term cataclasite can be applied to some of these rocks because of the degree of comminution. Some specimens are banded and yield evidence of possible flowage (planar alignment of sliver-like phacoidal clasts), but others contain randomly oriented fragments. Rocks are generally cemented by quartz and to a lesser degree by hematite. There is no evidence of any fault-related metamorphism as has been observed in some Cordilleran metamorphic core complexes (e.g., Rehrig and Reynolds, 1980). A zircon fission-track thermochronometer from the Su Casa area (sample NM-1) indicates that temperatures in that part of the lower plate have not exceeded ~175°C since ~59 Ma (Table 1).

It should also be noted that the microbreccia is thickest beneath subhorizontal parts of the NMDF and thinnest where the detachment steps down (increases dip to the northeast). It is speculated that the initial dip of the detachment may be the primary factor that controls the degree and location of lower-plate cataclasis.

Basement shear zones that parallel the local orientation of the NMDF also occur below the cataclastic zone. These zones are 0–1 m wide and are composed of anastomosing faults that border phacoid-shaped regions of cataclastic rock. Basement shear zones are best observed in the north-central part of the Newberry Mountains, north of the Azucar Mine.

Timing of movement on the NMDF can be constrained on the basis of well-dated cross-cutting relationships. The older limit is established by the age of the youngest rock that has been displaced along the NMDF. Nason and others (1979) presented a K-Ar date of 23.1 ±2.0 Ma on a dacite flow that lies in tectonic contact with the detachment in the northeastern Newberry Mountains. The younger limit is based on the age of the oldest postkinematic sedimentary rocks that rest directly on the detachment. Near Su Casa, the base of the middle Miocene Barstow Formation unconformably overlies the cataclastic lower-plate granite. The Barstow Formation was deposited between 16 and 13 Ma (Burke and others, 1982). These data constrain the time of movement along the NMDF to 23–16 Ma.

Breakaway Fault. The terms proximal and distal have been proposed to describe the positions of areas within the extended terranes (Dokka, 1983b). The majority of upper-plate normal faults dip toward the distal area of the terrane, whereas the tilted strata of upper-plate fault blocks dip in the direction of the proximal area. The terms breakaway or headwall fault have been applied by some workers (e.g., Howard and others, 1982) to the fault that separates the proximal area of the extended region from undeformed areas.

The breakaway fault of the Daggett terrane is presumed to be buried beneath Stoddard Valley and the western part of Kane Springs Wash and marks the southwestern limit of the terrane (Fig. 3). Areas southwest of the breakaway were apparently not extended during early Miocene time, judging from the lack of middle Tertiary extensional structures in the region. The nonextended area was a highland during and after deformation; it shed basement detritus into an evolving arcuate perimeter basin created along the headwall and the Kane Springs fault. Coarse, basement-derived clastics (formation of Slash X Ranch and the formation of Stoddard Valley) were deposited in this narrow structural basin.

Although the breakaway fault is not seen in outcrop because of burial by syntectonic deposits and younger rocks and sediments, its presence in Stoddard Valley is required because of the following field relations: (1) lower Miocene strata of the formation of Slash X Ranch dip steeply (50°) to the southwest (Fig. 3) and project into and beneath basement rocks that crop out on the

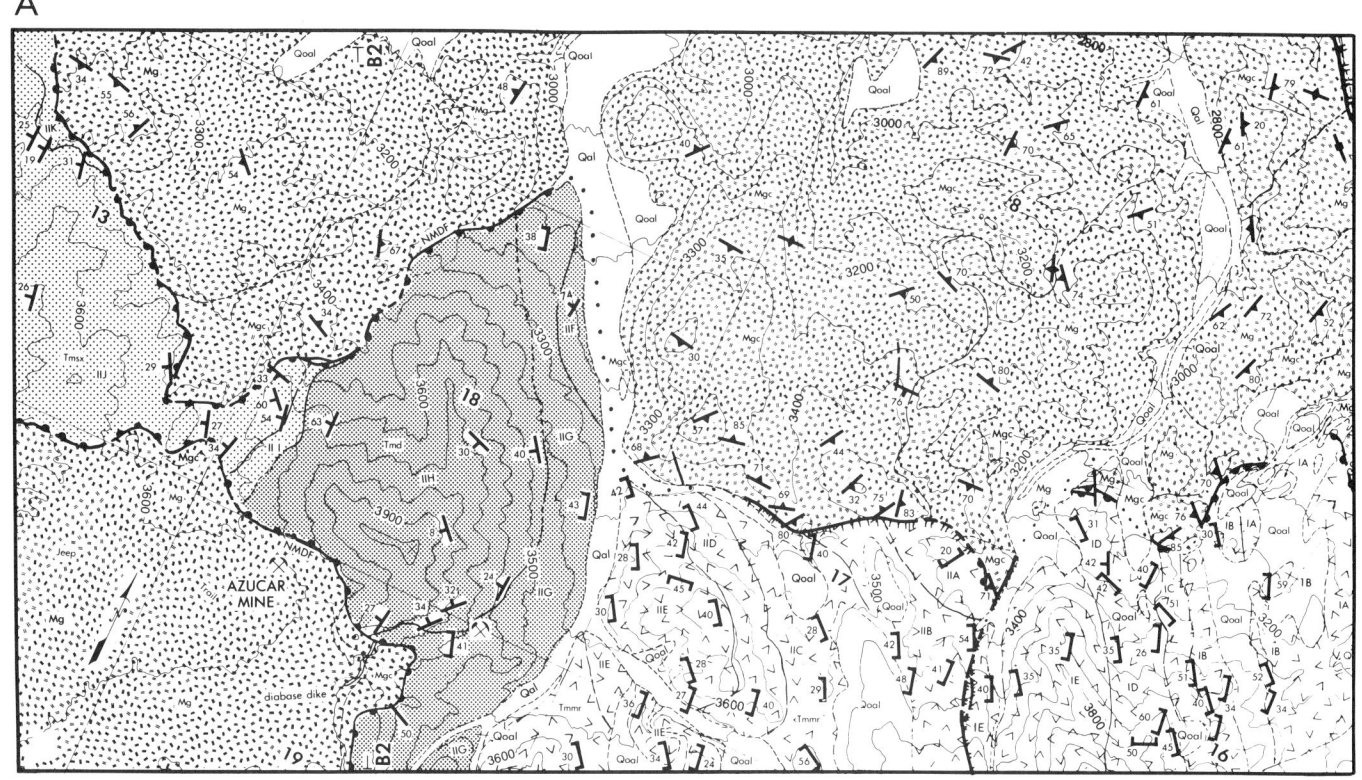

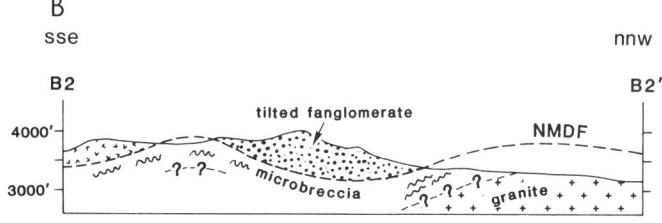

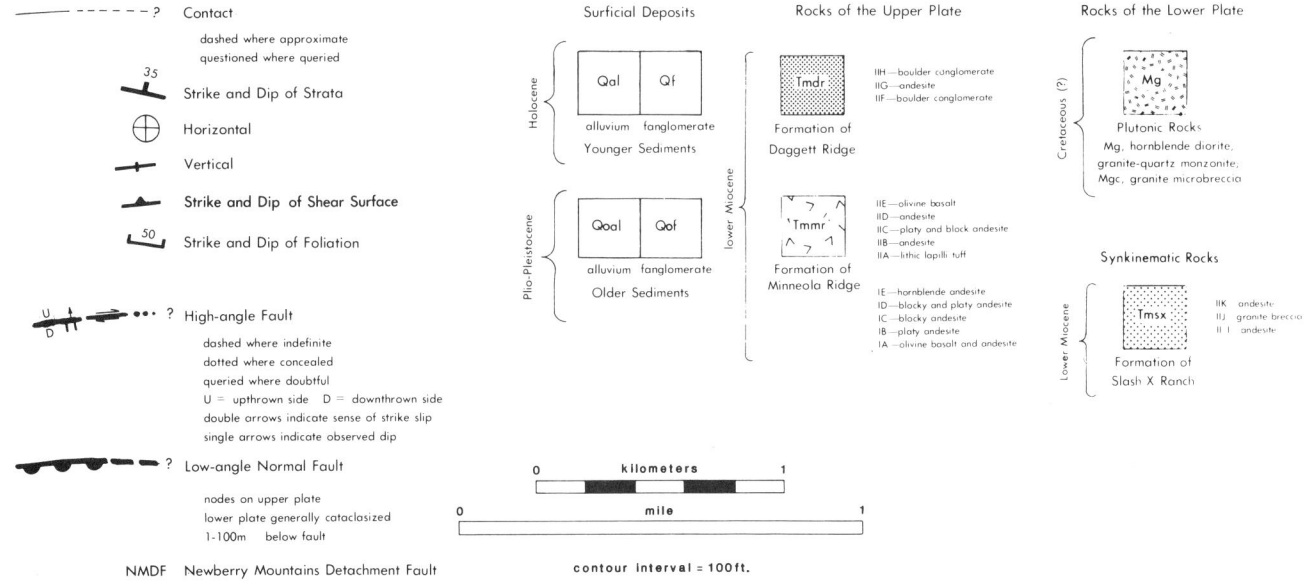

Figure 4. Geological map (A) and cross section (B; B2-B2′) of the Azucar Mine area, north-central Newberry Mountains illustrating the curviplanar form of the Newberry Mountains detachment fault (NMDF). Location of map shown on Figure 3. See text for discussion.

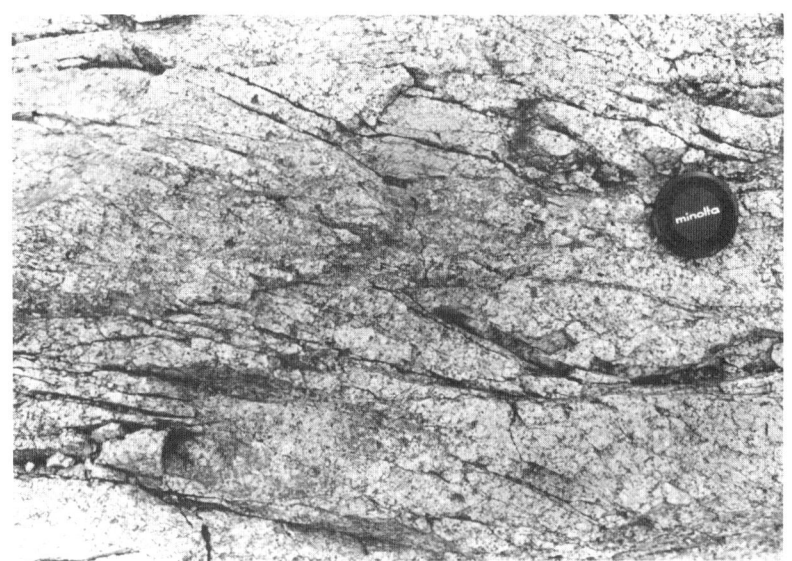

Figure 5. A, B, C: Outcrop photographs of a granite beneath the Newberry Mountains detachment fault (NMDF) in various stages of cataclasis. See text for discussion.

82   R. K. Dokka

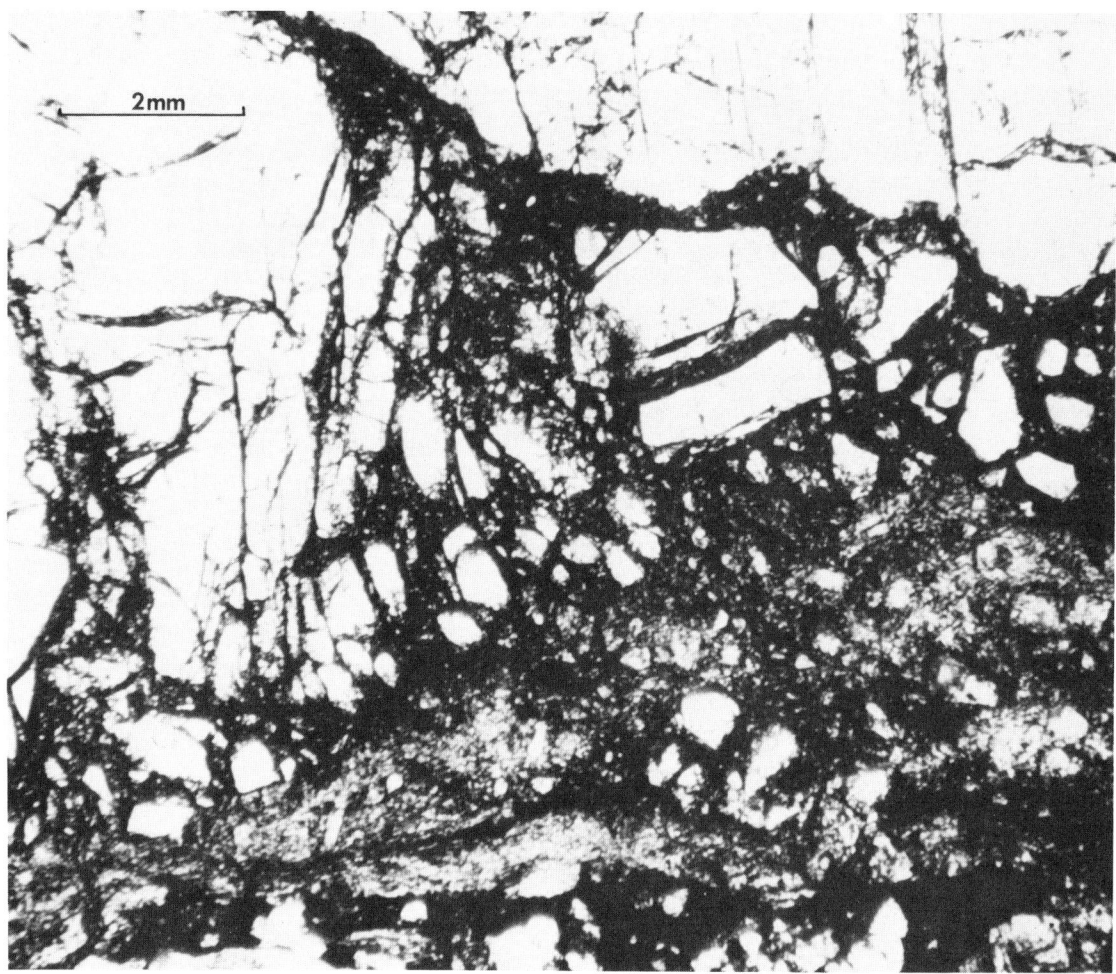

Figure 6. Photomicrograph of microbreccia shown in Figure 5C.

TABLE 1. FISSION-TRACK DATA

SAMPLE (latitude, longitude)	Rock unit	Mineral Dated	$P_s \times 10^6$ Tracks/cm^2 (counted)	$P_i \times 10^6$ Tracks/cm^2 (counted)	$\times 10^{15}$ Neutrons/cm^2 (counted)	No. of grains or fields	Etching Conditions	Age$\pm 1\sigma$(Ma)
NM-1 (34.784°, 116.874°)	Granite	Zircon	9.58 (345)	4.86 (175)	1.00 (2000)	4	NaOH-KOH 16h @220°C	58.7$\pm$5.6
NM-2 (34.725°, 116.744°)	Tuff of Kane Springs	Zircon	4.46 (375)	4.73 (397)	0.746 (2000)	9	NaOH-KOH 20h @220°C	21.0$\pm$1.6
ORD-1 (34.730°, 116.816°)	Granite	Zircon	4.73 (317)	2.52 (169)	0.840 (2000)	5	NaOH-KOH 18h @220°C	47.0$\pm$4.6
CM-4 (34.874°, 116.248°)	Tuff	Apatite	0.261 (470)	0.174 (314)	0.269 (2000)	82	7% HNO$_3$ 30s @20°C	24.0$\pm$2.6

southwest side of Stoddard Valley; (2) lower Miocene strata become progressively less steeply dipping upsection, suggesting hanging-wall sedimentary growth along an evolving normal fault (breakaway fault) located in Stoddard Valley (Fig. 3); and (3) these lower Miocene strata thin markedly to the northeast away from Stoddard Valley and consist of monolithologic landslide breccias and debris-flow conglomerates that had nearby sources to the southwest (Figs. 7 and 8) (Dokka, 1980).

The nature of hanging-wall deformation and associated sedimentation history provides important clues for determining the original shape and age of the breakaway fault. As shown on Figure 7b, lower Miocene strata of the formation of Slash X Ranch adjacent to the breakaway fault become progressively less steeply tilted upsection. The occurrence of syntectonic sedimentary deposits which show increasing dip with age (growth-fault deposits) is a strong argument that the breakaway fault was curviplanar and flattened with depth (Wernicke and Burchfiel, 1982). Hanging-wall deformation characteristic of curviplanar normal faults such as rollover antiforms and keystone grabens can also be observed. Activity during the deposition of the formation of Slash X Ranch is also indicated from the unit's depositional character. The composition and sedimentology of depositional units of the formation of Slash X Ranch are similar to ancient landslide deposits described by Heim (1932), Burchfiel (1966), Hsü (1975) and Krieger (1977), and are thought to indicate deposition along an active mountain front. These rocks rest above lower Miocene volcanic rocks and are in turn overlain by slightly younger sedimentary rocks and an intercalated ash-flow tuff (sample NM-2; 21.0 ±1.6 Ma) belonging to the formation of Stoddard Valley (Table 1) (see below for further discussion). Thus, the age of breakaway faulting can be constrained as between 23.1 and 21.0 Ma.

Upper Plate Fault Blocks. Upper-plate tilted fault blocks consist of Mesozoic crystalline basement rocks that are unconformably overlain by lower Miocene (24–21 Ma) silicic-mafic volcanic and sedimentary rocks. These blocks are bounded by a family of northwest- to northeast-striking, east-dipping, moderate-angle normal faults that display a consistent down-to-the-east sense of slip. The strike of these faults is generally north to north-northwest in the Newberry Mountains but progressively more northeasterly toward the east; faults strike from east-west to north to north-northeast in the Cady Mountains. Antithetic faults are rare.

A major consequence of this faulting is the repetition of the lower Miocene stratigraphic sequence across the entire terrane; this is best observed in the central Cady Mountains where faulting has disrupted a distinctive basal lower Miocene sequence (Fig. 9).

Analysis of kinematic indicators measured on the upper-plate faults suggests that movement was mainly dip-slip (Fig. 10). Slickensides striations and grooves measured on individual faults in the Newberry Mountains generally trend perpendicular to the strike of the faults (N50° ±10°E) (Fig. 10a) and are subparallel to the slip line of the NMDF. The trends of slip lines in the Cady Mountains were measured to be E-W ±30° (Fig. 10b); slip lines in this range pitch steeply (~75°) to the north, indicating a small component of left-slip (Mathis and Dokka, in prep.). In all areas, the trend of the local slip-line is perpendicular to the axis of rotation of tilted upper-plate strata.

Rotation of upper-plate rocks and structures also accompanied normal faulting. Upper-plate strata and crystalline rocks of the terrane are uniformly tilted ~40° (range 30°–70°) to the southwest (Fig. 2); observation of tilt patterns near the moderate-angle normal faults suggests that reverse drag was not common. Upper-plate normal faults may also have rotated by a similar amount, based on a comparison of original fault dip and postkinematic orientation. In order to determine the amount of rotation that a fault has undergone, it is first necessary to establish the original orientation of the fault. The initial dip can be inferred from the fracture angle (i.e., the acute angle between the fault and the adjacent bedding), if it is assumed that the strata were horizontal just prior to the time of faulting. This is a reasonable assumption because strata of the Daggett terrane were deposited just prior to faulting. For example, field measurements in the central Cady Mountains indicate that the moderate-angle normal faults were originally inclined 70°–85° to the northeast. The faults were subsequently tilted 30°–40° about an axis that was subparallel to the strike of the faults. Thus, the sense and amount of fault tilting was similar to the rotation recorded by the strata.

Well-dated cross-cutting relations indicate that these faults were active during early Miocene time; age limits are based on the age of the youngest rocks cut by the faults and on the age of untilted sedimentary and volcanic rocks that unconformably overlie the rotated, fault-bounded blocks. Post-tectonic strata unconformably overlie the tilted fault blocks in the central Cady Mountains and in the Newberry Mountains. Woodburne and others (1974) presented a K-Ar date of 21.6 ± 1.1 Ma (recalculated according to Dalrymple, 1979) on a tuff bed located near the base of the Hector Formation, which rests unconformably on a tilted lower Miocene volcanic sequence (Fig. 9). A tuff unit at the base of the tilted older section has been dated by the fission-track method on apatite as 24.0 ± 2.6 Ma (sample CM-4; Table 1). In the southern Newberry Mountains, the age of upper-plate faulting is constrained by the same cross-cutting relations used to constrain the time of movement along the breakaway fault (23–21 Ma). The interval of development of these upper-plate structures also overlapped the time of formation of the NMDF.

Mapping in the north-central Newberry Mountains has yielded an important clue to understanding the structural relationship of the upper-plate faults to the detachment and to the overall structure of the terrane. A geologic map and accompanying cross sections of the area (Fig. 11) illustrate the nonplanar nature of the NMDF, but more importantly show that the upper-plate faults do not displace the NMDF. Instead, these faults merge with the detachment at a low angle, thus requiring that the time of formation of the detachment be later or coincident with the extension of the upper plate. This can be resolved in the favor of a synchronous model by considering the following arguments: (1) the lower plate does not contain the truncated remains of

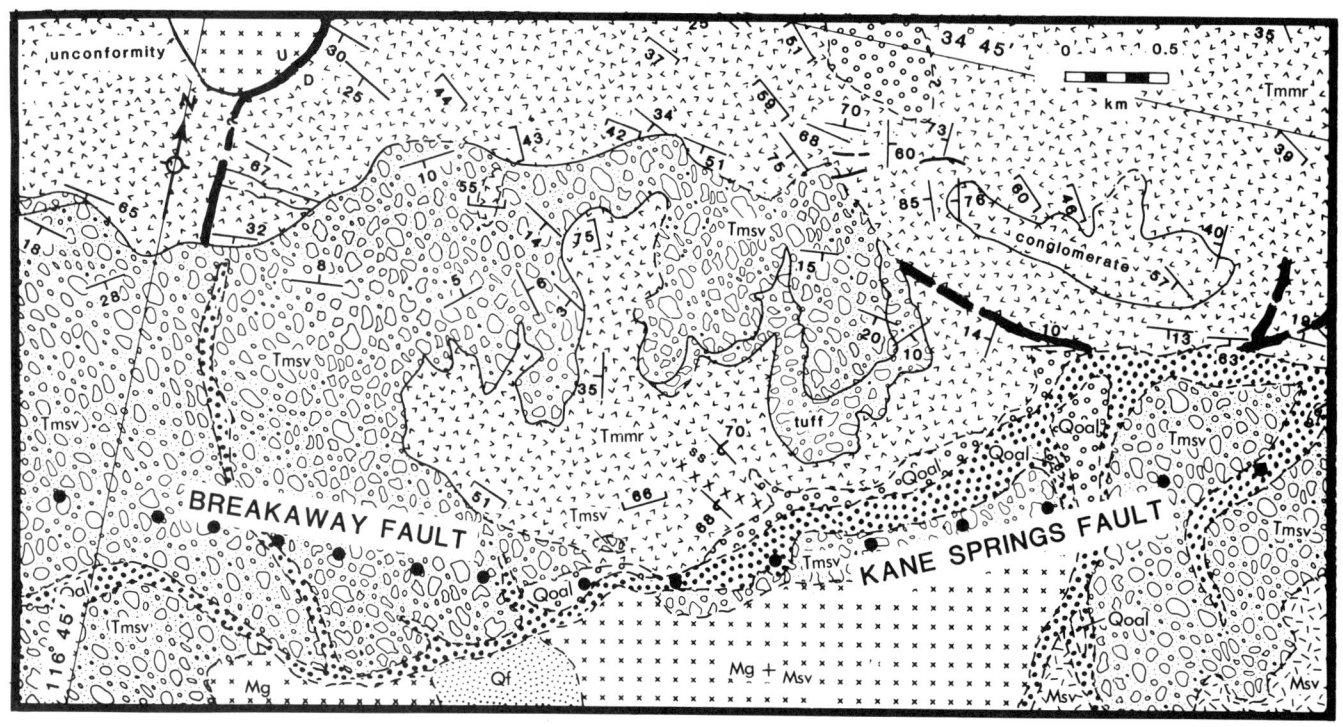
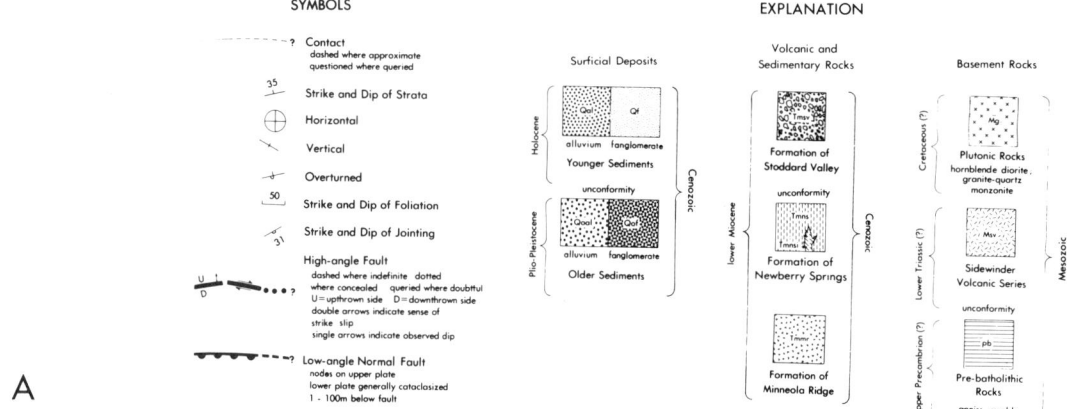
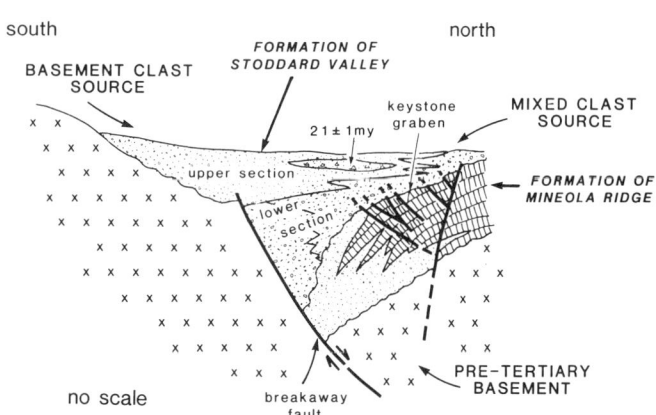

Figure 7. Geologic map (A) and idealized north-south cross section (B) of the Kane Springs area of the southern Newberry Mountains showing stratigraphic and temporal relations between syntectonic formation of Stoddard Valley and underlying tilted upper-plate rocks of the Daggett terrane. Location of map shown on Figure 3.

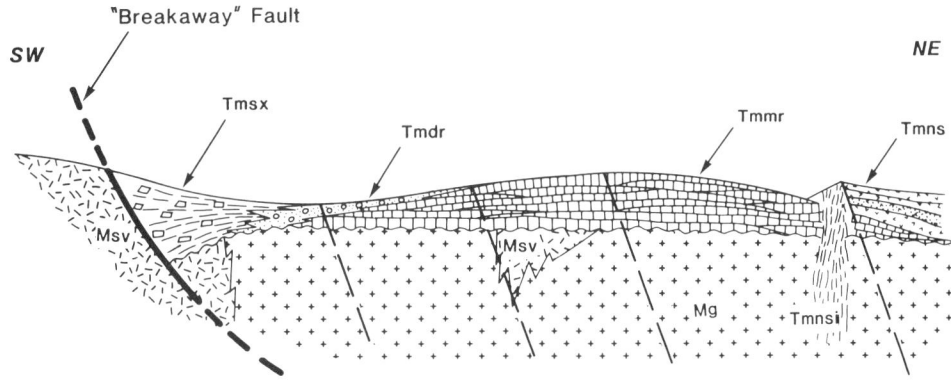

Figure 8. Restored section (ca. 21 Ma) of the proximal area of the Daggett terrane. Section oriented parallel to the major extension direction (southwest-northeast).

older high-angle normal faults as would be expected if the detachment was a late feature; and (2) upper-plate rocks (e.g., lower Miocene volcanic rocks) that are nose-down into the detachment (Fig. 4a) are never found in a lower-plate position. Synchronous and coordinated movement of upper-plate faults with the NMDF is also supported by kinematic measurements (Fig. 10).

Second Generation High-Angle Normal Faults. A younger family of high-angle normal faults has also been recognized in the Newberry Mountains (Dokka, 1980). This family is geometrically and kinematically similar to the older and more regionally developed group of faults. The faults strike north to northwest and dip moderately to the east; hanging-wall motion was down to the east. They truncate the older NMDF and associated tilted upper-plate rocks and structures (Fig. 12). Although one fault of this group has at least 2 km of dip-slip (Fig. 12), only minor (~5°) tilting was associated with this faulting event. Minor tilting and normal faulting of a similar orientation have also been noted in the northern Cady Mountains (Moseley, 1978; Miller, 1980; Williamson, 1980; Dokka, unpub. mapping) in association with sedimentary rocks deposited subsequent to the major tilting and faulting event.

Early Dike Swarm Emplacement. Prior to extension and rotation of the upper plate, parts of the Daggett terrane were intruded by localized dike swarms of andesite-dacite composition. Although these swarms contributed little to the overall extension of the region, they do represent zones in which there was concentrated strain. These structures are also important in that they allow us to infer the state of stress of that part of the crust at the time of intrusion. These intrusions are probably hypabyssal equivalents of the extrusives that built the Mojave Volcanic Belt. The largest dike swarms are located in the Newberry Mountains (Fig. 12) and the Cady Mountains (Fig. 9), and are oriented N20°W, 60°NE and N50°E, and 90°, respectively. Removal of the effects of rotation during subsequent normal faulting shows that the original dip of the dikes in both areas was approximately 90°. Dike swarms in both areas are mainly composed of parallel groups of tabular hypabyssal sheets, each typically 0–2 m thick. Observations on the geometry and kinematics of smaller dikes and of veins filled with fibrous minerals, which both occur in the crystalline basement adjoining each major dike complex, suggest that the initial fractures along which the dikes intruded were formed by extensional fracturing (i.e., fracturing in which there is no component of shear). The age of emplacement of the swarms can be inferred from the age of nearby petrologically similar lava flows. Data presented in Nason and others (1979) and Dokka (1980) for the Newberry area and by Woodburne and others (1974) for the western Cady Mountains indicate that dike emplacement occurred about ~24–23 Ma. The different emplacement orientations in the two areas suggest that the stress field beneath the central Mojave Desert area was nonuniform just prior to the major southwest-northeast extension. The width and orientation of the dike swarms suggest that the Newberry Mountains were dilated 2 km along a N70°E-trending horizontal line, whereas the central Cady Mountains were extended 0.5 km along a N40°W-trending horizontal line. Miller and others (1982) have also reported the occurrence of a dense swarm of northwest-striking dacite dikes in the eastern Bullion Mountains. The dikes constitute approximately half of a 9-km-wide outcrop area (100% extension along a horizontal axis).

Bullion Terrane

This terrane occurs south of the Daggett terrane and includes the central and southern parts of the Bullion Mountains (Fig. 2). Data regarding this terrane are limited because much of it lies within restricted areas of the U.S. Marine Corps Twentynine Palms Training Center. The data base consists of maps published by the U.S. Geological Survey (Dibblee, 1967b; Bassett and Kupfer, 1964) and the author's unpublished geologic mapping. Figure 13 is a geologic map of the northern Bullion Mountains showing parts of both the Bullion and Daggett terranes, as well as a strike-slip fault (Kane Springs fault) that separates them.

Faults of the Bullion terrane are generally of opposite polar-

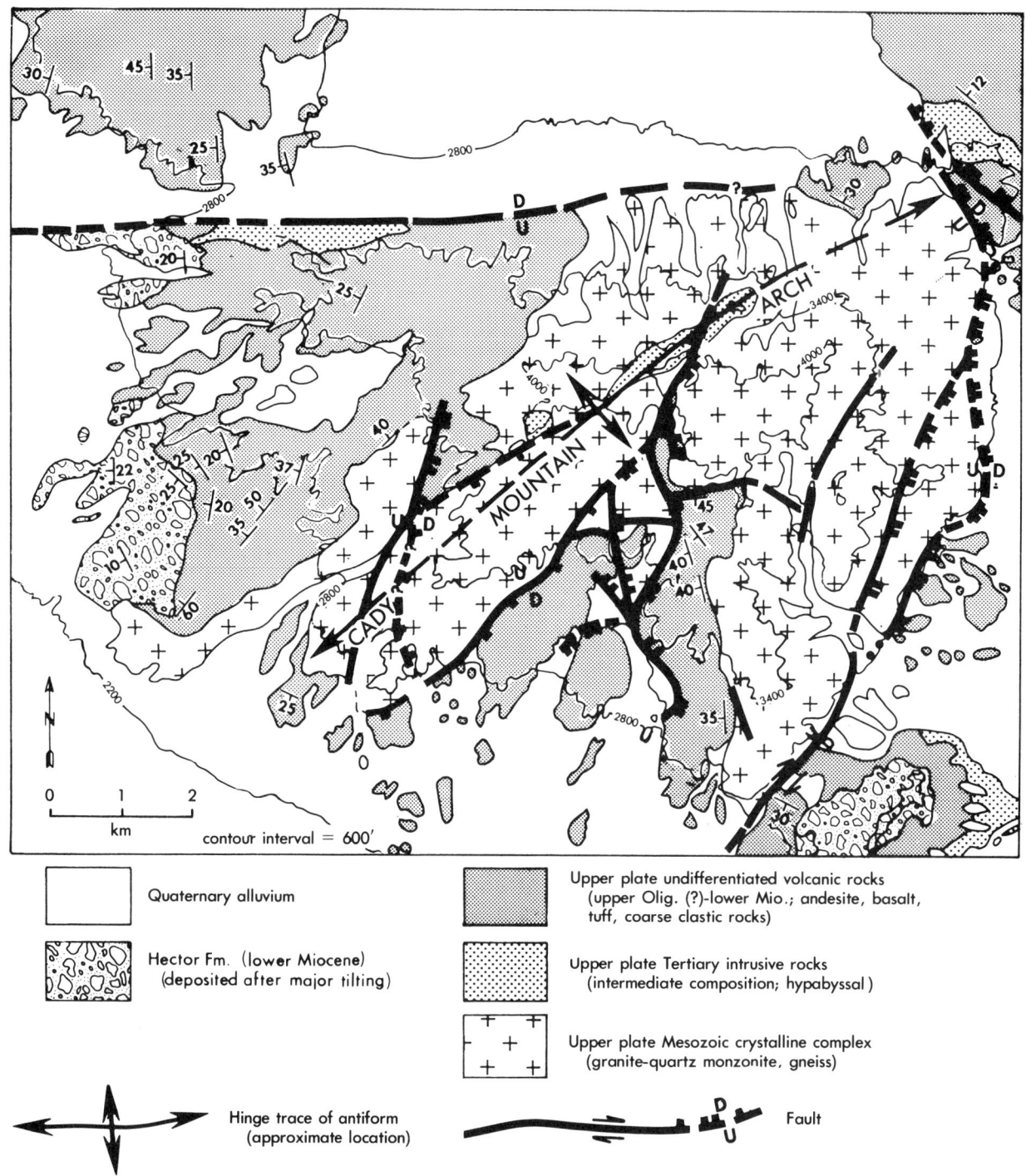

Figure 9. Geologic sketch map of the central Cady Mountains.

ity (dip to the southwest) and dip more steeply (55°–65°) than do the upper-plate faults of the Daggett terrane. Strata dip less (20°–40°) and in an opposite direction (east-northeast) to the beds of the Daggett terrane. Limited measurements of fracture angles suggest that the faults originally dipped steeply (~85°) to the southwest. The results of a geometric analysis of tilting suggest that both rocks and structures of the Bullion terrane have been rotated 20°–40° about a subhorizontal, northwest-trending axis. This rotational axis is similar in orientation but opposite in sense, to the rotational axis determined for the Daggett terrane.

Kane Springs Fault

Because of its vertical and lateral inhomogeneity, the lithosphere must extend nonuniformly in terms of strain distribution and strain rates (Dokka and Pilger, 1983). Therefore, it is not surprising that areas of continental extension such as the CMEC contain internal transform fault zones that served to accommodate differential lateral extension. Such strain boundaries are critical paleotectonic elements in the structure of an extension orogen.

Although now disrupted by late Cenozoic strike-slip faulting, the trace of the Kane Springs fault can be located in each of the ranges from the western Newberry Mountains to the Ludlow area (Fig. 14). This fault was named by the author on the basis of relationships along Kane Springs Wash in the southern Newberry Mountains area (Dokka, 1980). The Kane Springs fault is also important as a regional structure that has been used as a marker to establish the magnitude and sense of post-20 Ma right shear across the central Mojave Desert (Dokka, 1983a).

The Kane Springs fault displays strike slip and oblique slip along different parts of its trace. Segments of the fault located east of Kane Springs (Fig. 14) are subvertical, strike N70°E, and yield subhorizontal kinematic indicators. In these ranges the fault is an anastomosing zone that has produced a linear zone of cataclastic rocks and gouge (Fig. 15). The fault zone is widest in the northern Bullion Mountains where it separates the two terranes (Fig. 13) and is narrowest in the Rodman Mountains where it lies between the Daggett terrane and an unextended region. Although kinematic data (Fig. 10c) indicate that this segment is conservative (slip subparallel to fault strike), topographic relief was created along the fault as a consequence of the extension and associated subsidence of the Daggett terrane. Because deformation was apparently concentrated in the crust, the extended terranes tended to subside isostatically (e.g., McKenzie, 1978). This was in marked contrast to the unextended area south of the fault which remained high throughout tectonism. This highland served as the source for alluvial fans that built northward onto rotating fault blocks (Dokka and others, in prep.).

From near Kane Springs west to Stoddard Valley (Fig. 14), the trace of the Kane Springs fault gradually becomes more northwesterly (i.e., changes clockwise to a west-northwest orientation). A narrow band of syn- and post-kinematic sedimentary rocks occurs along this segment of the fault. This unit has been informally referred to as the formation of Stoddard Valley (Dokka, 1980) and has been subdivided into two parts: (1) a lower unit of unorganized matrix-supported boulder conglomerate and breccia; and (2) an upper unit of relatively flat-lying, well-bedded sandstone, conglomerate, and 21 Ma tuff. These latter rocks unconformably overlie the chaotic lower unit and steeply tilted lower Miocene volcanic rocks belonging to the upper plate of the Daggett terrane (Fig. 7). The lower unit of the formation of Stoddard Valley is interpreted to be the result of very rapid deposition of basement detritus into an evolving basin, whereas the upper flat-lying unit is considered to represent post-tectonic basin fill and reworked older sediments.

The origin of the basin in which the formation of Stoddard Valley accumulated can be explained as being the result of changes in the geometry and orientation of the Kane Springs fault

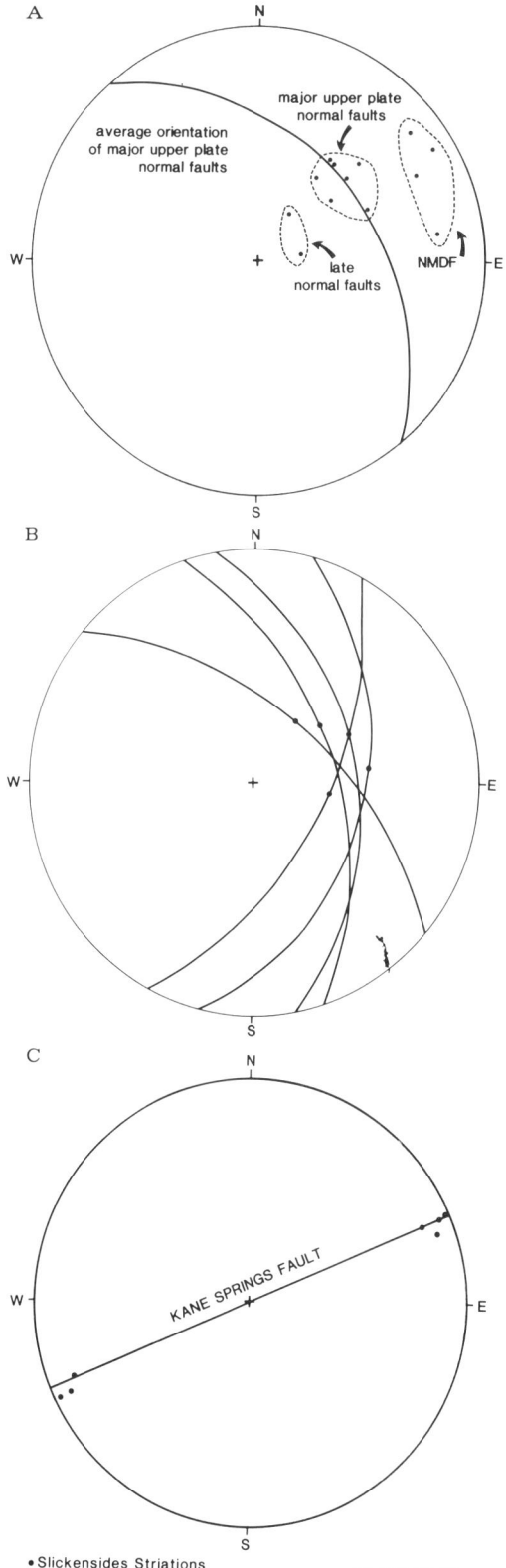

Figure 10. Summary diagrams of kinematic indicators from areas within the Central Mojave extensional complex (CMEC). A: Newberry Mountains; B: Cady Mountains; (C): Kane Springs fault (southeastern Newberry Mountains and Rodman Mountains).

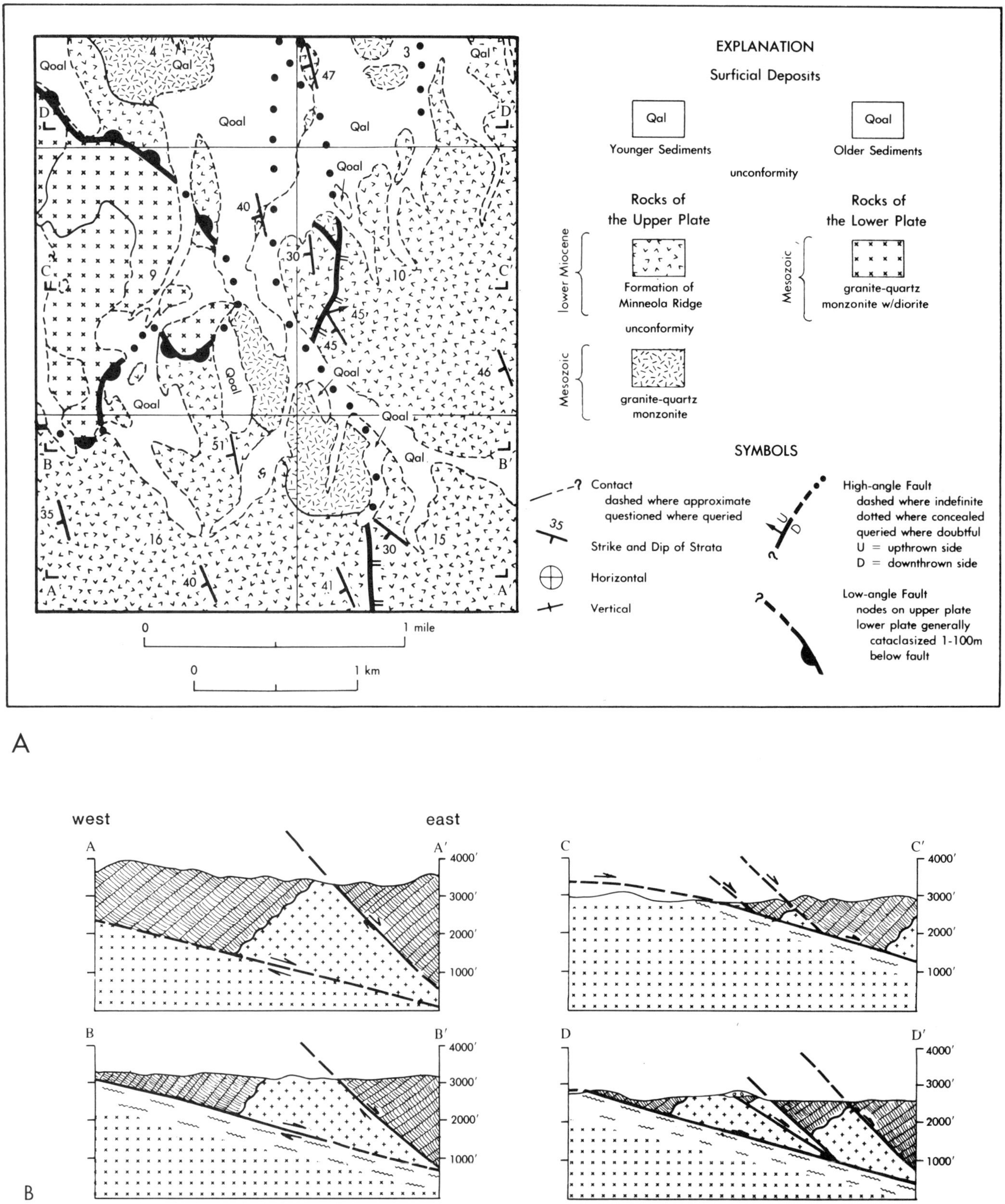

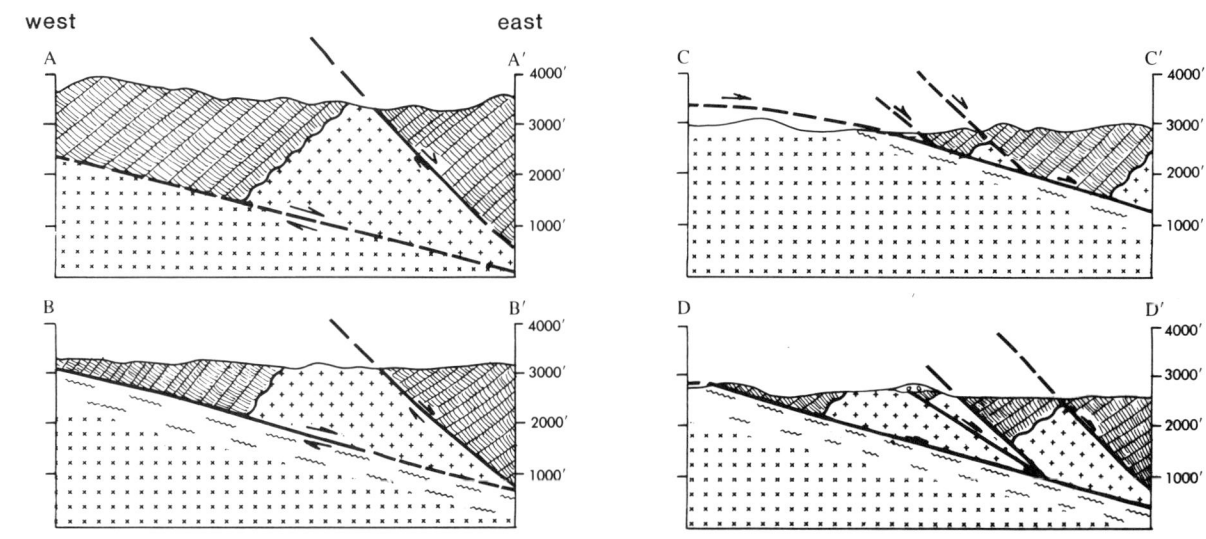

Figure 11. Geologic map (A) and (B) cross sections (A–D) of the north-central Newberry Mountains.

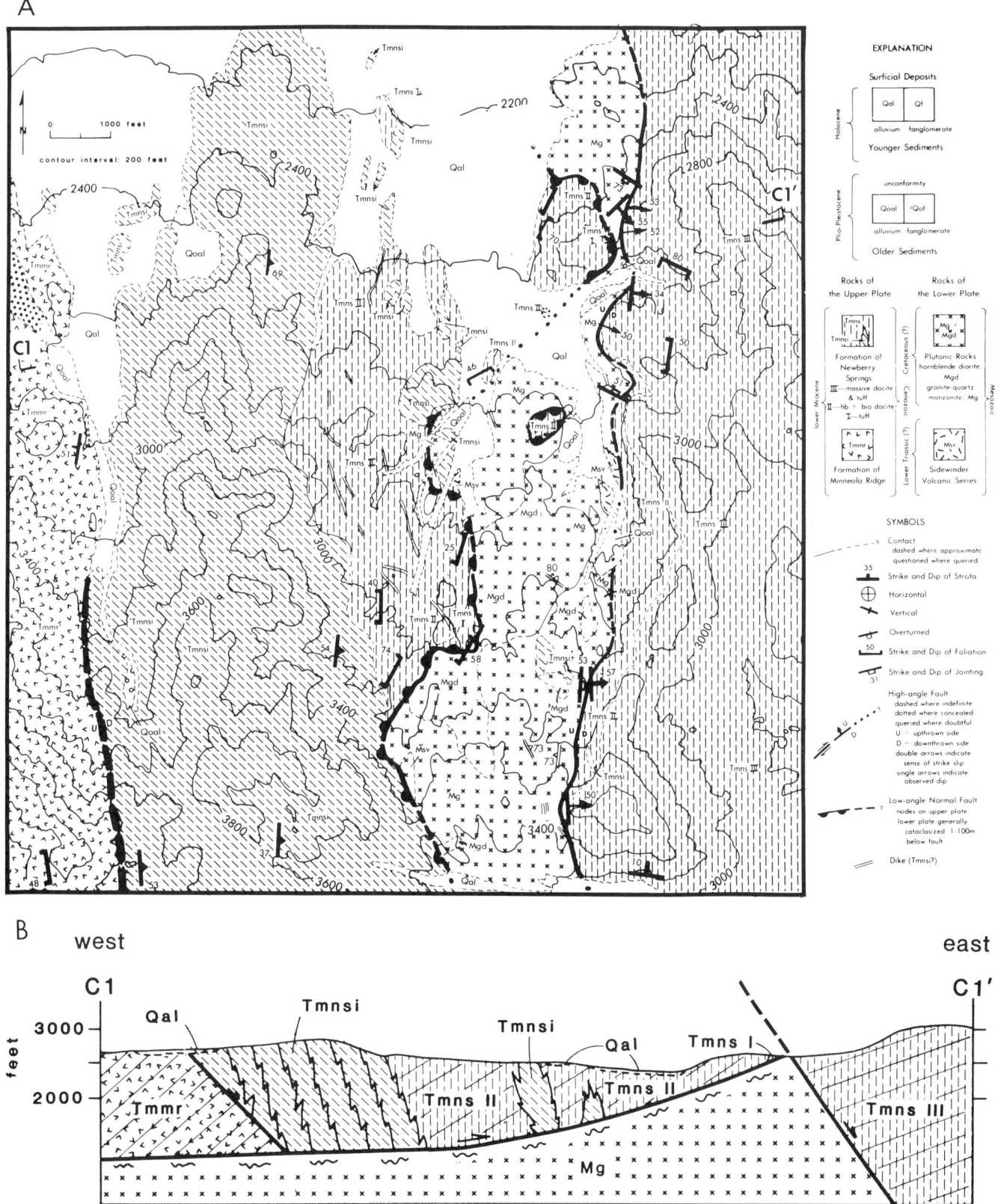

Figure 12. Geologic map (A) and cross section (B) of the Newberry Springs area showing a second generation high-angle normal fault that cuts and displaces the Newberry Mountains detachment fault (NMDF); low-angle normal fault). Location of map shown on Figure 3.

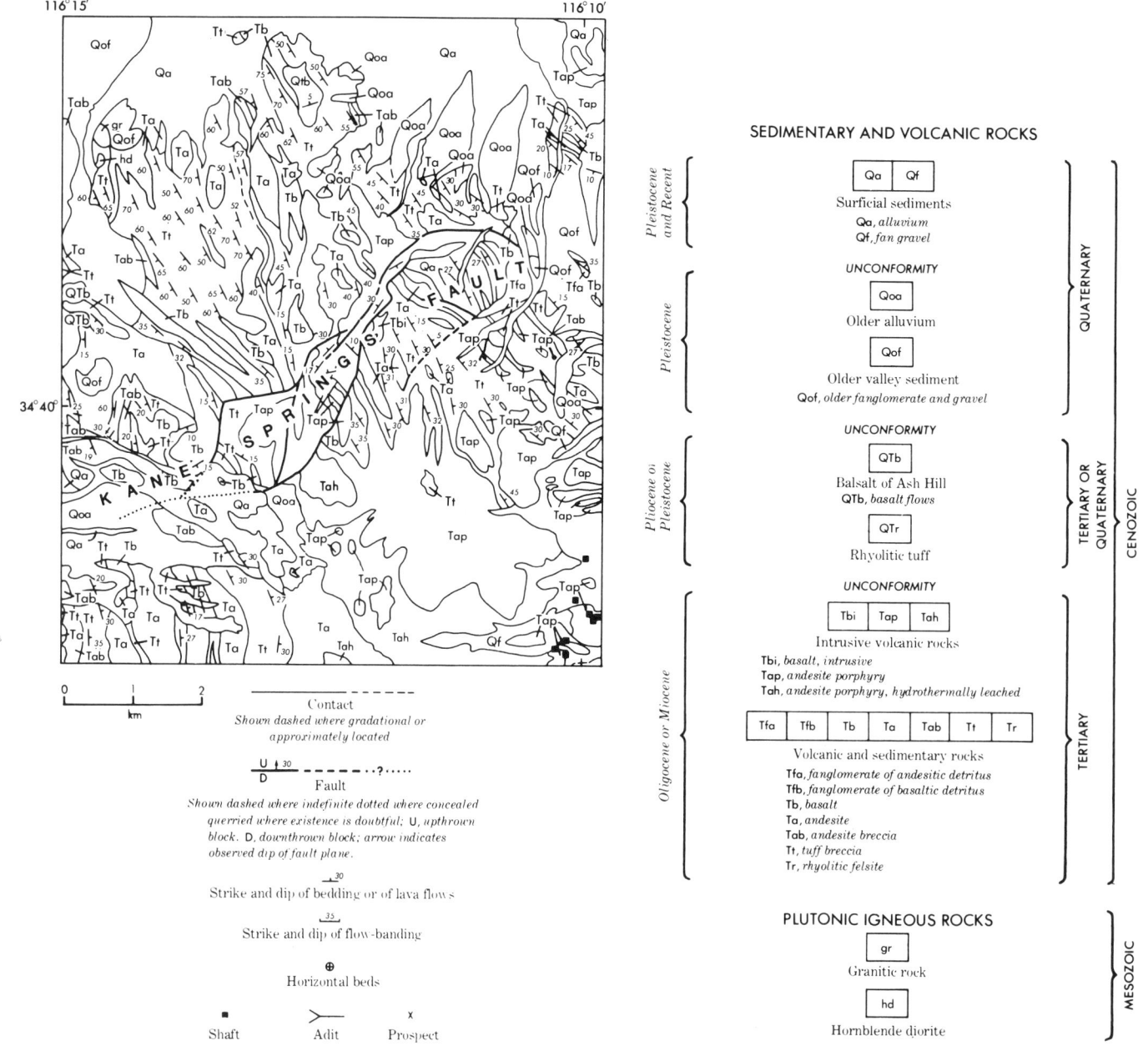

Figure 13. Geologic map of the northern Bullion Mountains (modified from Dibblee, 1967b) showing juxtaposition of the Daggett (north) and Bullion (south) terranes along the Kane Springs fault.

along its strike. Measurement of fault-plane kinematic indicators, such as slickensides striations, grooves, and fibrous mineral growths, were collected along the length of the fault (Fig. 10c). Data from the segment east of the central Rodman Mountains indicate that motion was mainly subhorizontal and occurred along a ~N60°E-trending line. This slip-line is subparallel to both the local strike of the Kane Springs fault and the horizontal extension axis determined for both of the adjacent extended terranes. Thus, data suggest that this segment of the fault was dominantly of strike-slip character. West of the central Rodman Mountains, the strike of the Kane Springs fault gradually changes clockwise to the west-northwest; the horizontal extension axis of the adjacent Daggett terrane remains oriented ~N60°E. This nonparallelism of extension axis and fault strike requires that movement along this segment be oblique, consisting of both strike-slip and normal components (Fig. 16).

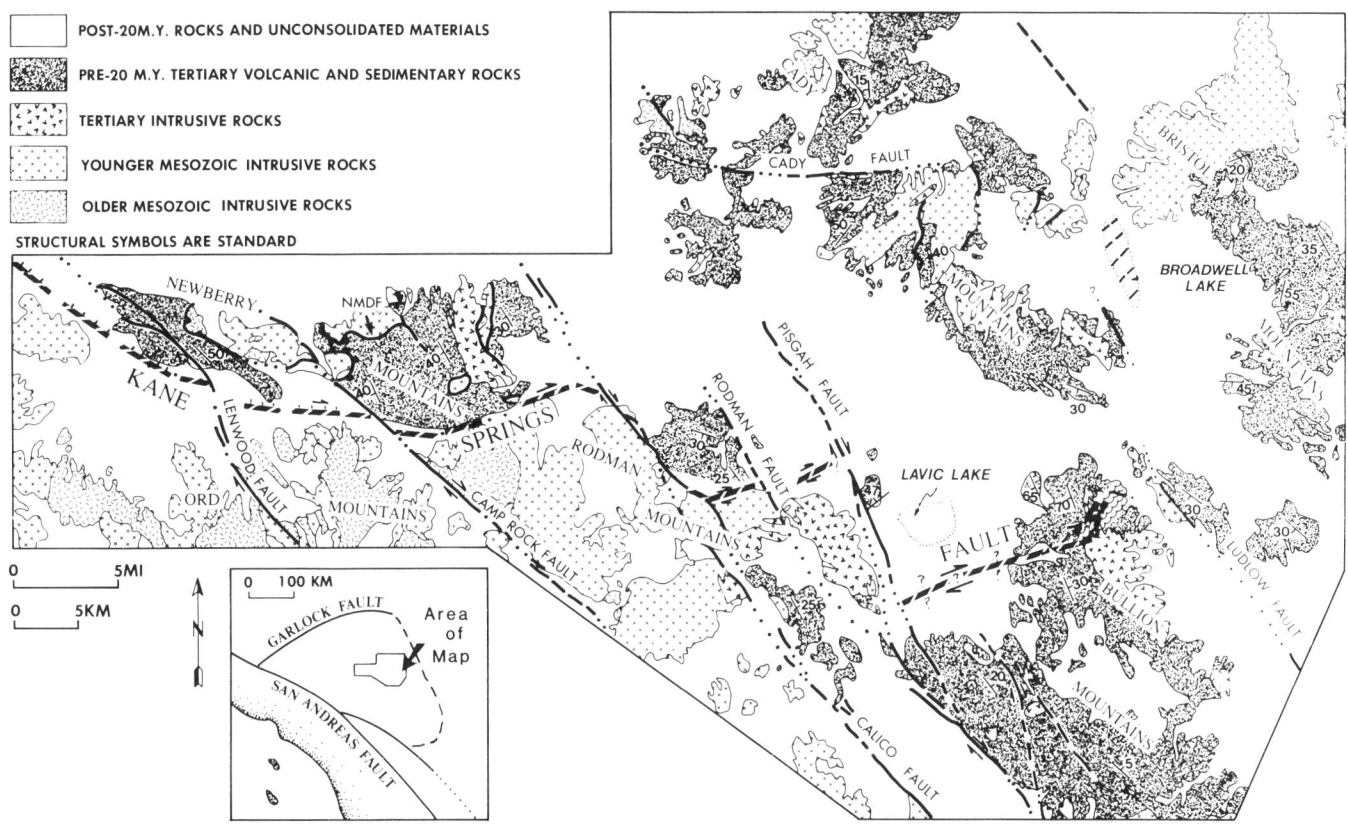

Figure 14. Generalized geologic map of the central Mojave Desert highlighting the now disrupted (post-20-Ma right-slip faulting) trace of the Kane Springs fault.

ARCHITECTURE OF EARLY MIOCENE EXTENSION OF THE CENTRAL MOJAVE DESERT

The overall geometry and lateral distribution of structures in the CMEC suggest that extension developed in at least two half-grabens (Bally, 1981); each of the aforementioned CMEC domains corresponds to a half-graben. According to Bally, a half-graben is a zone of extension dominated by a family of normal faults of similar orientation. The two half-grabens of the CMEC have opposite polarities and are separated by a synkinematic strike-slip fault, the Kane Springs fault. This pattern of separate but kinematically linked zones of extension is similar to oceanic networks of spreading ridges and transforms.

The key internal structural elements within an individual half-graben in the CMEC are: (1) an upper plate of brittlely extended and uniformly rotated normal-fault bounded blocks; (2) a downward-flattening breakaway fault; and (3) a curviplanar, shallowly dipping normal fault (detachment) that separates the extended upper plate from an undeformed lower plate. The region above the detachment was extended by a family of planar high-angle normal faults. Rotation of these faults to shallower orientations occurred in concert with tilting of upper-plate fault blocks and with motion along the detachment. The uniformity of tilting of upper-plate rocks across the half-graben strongly argues that upper-plate, high-angle normal faults were planar. This fault geometry is in marked contrast to the bounding curviplanar breakaway fault. The breakaway fault is considered to have a scoop-shaped geometry because of the development of reverse drag-related features (Hamblin, 1965) in the hanging wall and by the fault's curvilinear trace. Based on the fault's downward-flattening geometry and temporal coincidence with the detachment, it is speculated that the breakaway fault may be the uppermost part of the detachment. Figure 17 depicts the author's interpretation of the relationship between upper-plate faults, the breakaway fault, and the NMDF.

The geometry and kinematics of the structures within a CMEC half-graben are strikingly similar to those found in many Tertiary extensional complexes of western North America (e.g., Anderson, 1971; Davis and Coney, 1979; Davis and others, 1980; Carr and others, 1980; Wernicke, 1981; Allmendinger and others, 1983). A particularly puzzling aspect of these complexes has been that although lower-plate rocks vary in terms of their petrologic, thermal, and structural histories, the style of upper-plate extension is similar. Wernicke (1981) has proposed a unifying theory that attempts to relate these disparate data in terms of a rooted detachment model. I consider that this model can be

Figure 15. A: Oblique aerial photograph of the Kane Springs fault in the Rodman Mountains (view is to southwest). The fault separates unextended Mesozoic granitic rocks from uniformly tilted upper-plate volcanic rocks and syntectonic growth-fault deposits. B: Vertical aerial photograph of the same segment of the Kane Springs fault in the Rodman Mountains (top of photo is to northeast).

successfully used to explain the structural relations in the CMEC because it is the only current model that can account for the lack of penetrative extension and thermal disturbance beneath the lower plate. According to Wernicke's model, a detachment is a crustal-scale zone of simple shear that can play a major role in the rifting of the lithosphere. The observed regional variation in upper-plate–lower-plate relations is explained in terms of: (1) the amount of extension in a given detachment complex; and/or (2) the structural level of the detachment that is under observation (Fig. 18). Because the observed lower plate in the Daggett terrane is not mylonitic or metamorphosed, it is proposed that the half-grabens of the CMEC represent the near-surface part of a rooted detachment (Fig. 18).

CONCLUSIONS

The following conclusions were reached regarding the nature of early Miocene extension within the central Mojave Desert area:

1. The processes responsible for the observed extension include low-angle normal faulting, high-angle normal faulting, and extension fracturing. Extension fracturing (and associated dike

Figure 15C: Outcrop photograph of the Kane Springs fault taken near the mouth of Kane Springs Wash. Note the anastomosing character of the high-angle shear fractures that comprise this fault.

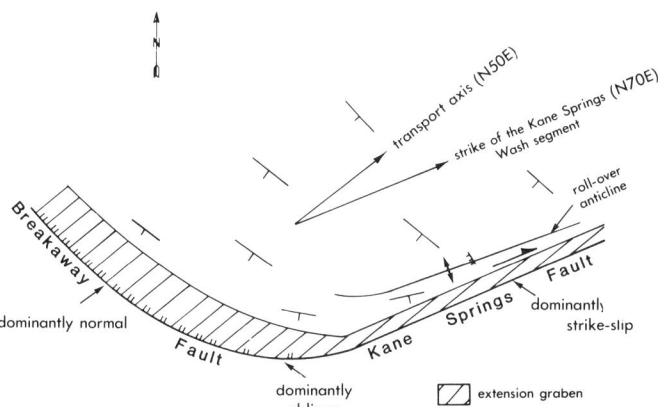

Figure 16. The relationship of fault geometry to the development of oblique extension along the eastern part of the breakaway fault.

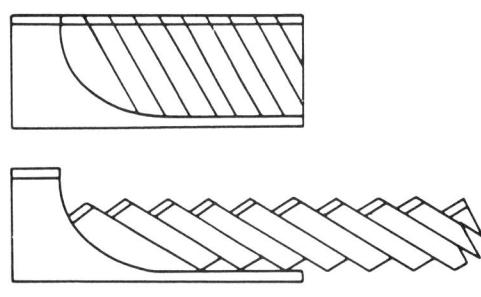

Figure 17. Inferred geometric relations between upper-plate normal faults, breakaway fault, and Newberry Mountains detachment fault (model from Wernicke and Burchfiel, 1982).

emplacement) developed early (~24–23 Ma), but only resulted in small, local dilations of the crust. This process was extremely important, however, in that it produced paths to the surface for magmas that formed the Mojave Volcanic Belt. Primary extension of the CMEC occurred slightly later and was accomplished by detachment faulting (coordinated low-angle normal faulting and upper-plate, high-angle normal faulting).

2. Extensional elements now exposed in the CMEC developed in the brittle field. There is no evidence for any first-order ductile deformation as has been observed in other Cordilleran extensional zones. This does not, however, rule out the possibility that structures developed by ductile processes occur at greater depth or in areas now covered by postextension strata or sediments.

3. The major phase of extension was concentrated in at least two domains that were kinematically linked by a transform fault (Kane Springs fault). The half-graben concept of Bally (1981) can be applied to describe the lateral arrangement of near-surface structural elements in each domain.

4. The internal geometry of one of the half-grabens consists

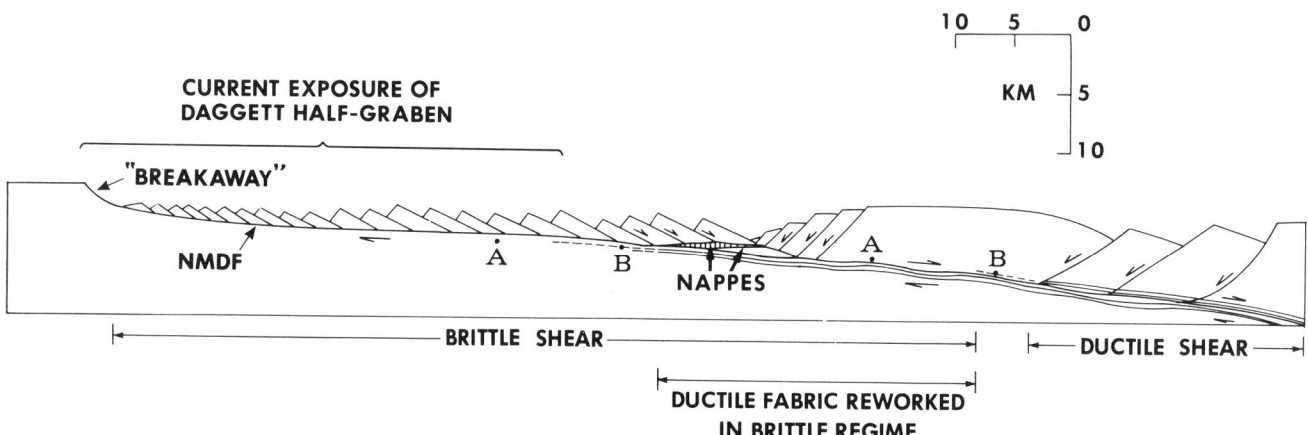

Figure 18. Rooted detachment model of Wernicke (1981). Proposed position of the Daggett half-graben in this model is indicated.

of: (a) a family of similarly rotated upper-plate blocks bounded by planar normal faults; (b) a scoop-shaped breakaway fault; and (c) a shallowly dipping, low-angle normal fault that floors the rotated and extended upper plate; the breakaway fault may be the near-surface expression of the detachment. Although the nature of the merger between upper-plate normal faults with the detachment is not completely understood, temporal and kinematic relations in the Daggett "half-graben" suggest that extension and rotation of upper-plate strata and crystalline rocks occurred in concert with movement along the NMDF. The structural association and the lateral and vertical arrangement of structures are consistent with the rooted detachment model proposed by Wernicke (1981).

REFERENCES CITED

Allmendinger, R. W., Sharp, J. W., Von Tish, D., Serpa, L., Brown, L., Kaufman, S., Oliver, J., and Smith, R. B., 1983, Cenozoic and Mesozoic structure of the eastern Basin and Range province, Utah, from COCORP seismic-reflection data: Geology, v. 11, p. 532–536.

Anderson, R. E., 1971, Thin skin distension in Tertiary rocks of southeastern Nevada: Geological Society of America Bulletin, v. 82, p. 43–58.

Armstrong, R. L., and Higgins, R., 1973, K-Ar dating of the beginning of Tertiary volcanism in the Mojave Desert, California: Geological Society of America Bulletin, v. 84, p. 1095–1100.

Bally, A. W., 1981, Atlantic-type margins, in Bally, A. W., ed., Geology of passive continental margins: History, structure and sedimentologic record: American Association of Petroleum Geologists Education Course Note Series 19, p. 1–48.

Bassett, A. M., and Kupfer, D. H., Jr., 1964, A geologic reconnaissance in the southeastern Mojave Desert, California: California Division of Mines and Geology Special Report 83, 43 p.

Burchfiel, B. C., 1966, Tin Mountain landslide, southeastern California, and the origin of megabreccia: Geological Society of America Bulletin, v. 77, p. 95–100.

Burchfiel, B. C., and Davis, G. A., 1980, Mojave Desert and surrounding environs, in Ernst, W. G., ed., The geotectonic development of California: Englewood Cliffs, New Jersey, Prentice-Hall, p. 217–252.

Burke, D. B., Hillhouse, J. W., McKee, E. H., Miller, S. T., and Morton, J. L., 1982, Cenozoic rocks in the Barstow Basin area of southern California - Stratigraphic relations, radiometric ages, and paleomagnetism: U.S. Geological Survey Bulletin 1529-E, 16 p.

Carr, W. J., Dickey, D. D., and Quinlivan, W. D., 1980, Geologic map of the Vidal NW, Vidal Junction, and parts of the Savahia Peak SW and Savahia Peak quadrangles, San Bernardino County, California: U.S. Geologic Survey Map I-1126, scale 1:24,000.

Dalrymple, G. Brent, 1979, Critical tables for conversion of K-Ar ages from old to new constants: Geology, v. 7, p. 558–560.

Davis, G. A., Anderson, J. L., Frost, E. G., and Shackelford, T. J., 1980, Mylonitization and detachment faulting in the Whipple-Buckskin-Rawhide Mountains terrane, southeastern California and western Arizona, in Crittenden, M. D., Coney, P. J., and Davis, G. H., eds., Cordilleran metamorphic core complexes: Geological Society of America Memoir 153, p. 79–130.

Davis, G. H., and Coney, P. J., 1979, Geologic development of Cordilleran metamorphic core complexes: Geology, v. 7, p. 120–124.

Dibblee, T. W., 1961, Evidence of strike-slip movement on northwest-trending faults in the Mojave Desert, California: U.S. Geological Survey Professional Paper 424-B, p. B197–B199.

—— 1967a, Areal geology of the western Mojave Desert, California: U.S. Geological Survey Professional Paper 522, 153 p.

—— 1967b, Geologic map of the Ludlow quadrangle, San Bernardino County, California: U.S. Geological Survey Miscellaneous Geologic Map Investigations I-477, scale 1:62,500.

—— 1970, Geologic map of the Daggett quadrangle, San Bernardino County, California: U.S. Geological Survey Miscellaneous Geologic Map Investigations I-592, scale 1:62,500.

—— 1971, A great middle Tertiary buttress unconformity in the Newberry Mountains, Mojave Desert, California, and its paleogeologic implications: Geological Society of America Abstracts with Programs, v. 3, p. 110.

Dokka, Roy K., 1979, Styles and timing of late Cenozoic faulting, central Mojave Desert, California: Geological Society of America Abstracts with Programs, v. 11, p. 414.

—— 1980, Late Cenozoic tectonics of the central Mojave Desert, California [Ph.D. thesis]: Los Angeles, University of Southern California, 220 p.

—— 1983a, Displacements on late Cenozoic strike-slip faults of the central Mojave Desert, California: Geology, v. 11, p. 305–308.

—— 1983b, Upper crustal deformation processes and strain transitions in an extensional orogen: Geological Society of America Abstracts with Programs,

v. 15, p. 287.

Dokka, R. K., and Pilger, R. H., Jr., 1983, A non-uniform extension model for continental rifting: Geological Society of America Abstracts with Programs, v. 15, p. 559.

Garfunkel, Z., 1974, Model for the late Cenozoic tectonic history of the Mojave Desert and its relation to adjacent areas: Geological Society of America Bulletin, v. 85, p. 1931–1944.

Hamblin, W. K., 1965, Origin of "reverse drag" on the downthrown side of normal faults: Geological Society of America Bulletin, v. 76, p. 1145–1164.

Heim, A., 1932, Bergsturz and Menscheleben: Zurich, Fretz Wasmuth Verlag, 218 p.

Hewett, D. F., 1954, General geology of the Mojave Desert region, California, *in* Jahns, R., ed., Geology of southern California: California Division of Mines Bulletin 170, p. 15–18.

Higgins, M. V., 1971, Cataclastic rocks: U.S. Geological Survey Professional Paper 687, 97 p.

Howard, K., Stone, P., Pernokos, M., and Marvin, R., 1982, Geologic and geochronologic reconnaissance of the Turtle Mountains area, California: West border of the Whipple Mountains detachment terrane, *in* Frost, E. G., and Martin, D., eds., Mesozoic-Cenozoic tectonic evolution of the Colorado River region, California, Arizona, and Nevada: San Diego, Cordilleran Publishers, p. 341–354.

Hsü, K. J., 1975, Catastrophic debris streams (sturzstroms) generated rock falls: Geological Society of America Bulletin, v. 86, p. 129–140.

Kistler, R., 1974, Phanerozoic batholiths in western North America: A summary of recent work on variations in time, space, chemistry, and isotopic compositions: Annual Reviews of Earth and Planetary Science, v. 2, p. 403–418.

Krieger, M. H., 1977, Large landslides, composed of megabreccia, interbedded in Miocene basin deposits, southeastern Arizona: U.S. Geological Survey Professional Paper 1008, 25 p.

McKenzie, D. P., 1978, Some remarks on the development of sedimentary basins: Earth and Planetary Science Letters, v. 40, p. 25–32.

Miller, D., Howard, K., and John, B., 1982, Geology of the Bristol Lake region, Mojave Desert, California, *in* Cooper, J. D., ed., Geologic excursions in the California desert (Geological Society of America Cordilleran Section Meeting guidebook): Anaheim, California, Geological Society of America Cordilleran Section, p. 91–100.

Miller, E., 1977, Geology of the Victorville region, California [Ph.D. thesis]: Houston, Rice University, 226 p.

Miller, S. T., 1980, Geology and mammalian biostratigraphy of a part of the northern Cady Mountains, California: U.S. Geological Survey Open-File Report 80-978, 121 p.

Moseley, C. G., 1978, Geology of a portion of the northern Cady Mountains, Mojave Desert, California [M.S. thesis]: Riverside, University of California, 131 p.

Nason, G. W., Davis, T. E., and Stull, R. J., 1979, Cenozoic volcanism in the Newberry Mountains, San Bernardino County, California, *in* Armentrout, J. M., Cole, M. R., and TerBest, H., Jr., eds., Cenozoic paleogeography of the western United States: Society of Economic Paleontologists and Mineralogists, Pacific Coast Paleogeography Symposium 3, p. 89–96.

Rehrig, W., and Reynolds, S., 1980, Geologic and geochronologic reconnaissance of a northwest-trending zone of metamorphic core complexes in southern Arizona, *in* Crittenden, M. D., Coney, P. J., and Davis, G. H., eds., Cordilleran metamorphic core complexes: Geological Society of America Memoir 153, p. 131–157.

Spencer, J., 1984, Role of tectonic denudation in warping and uplift of low-angle normal faults: Geology, v. 12, p. 95–98.

Stewart, J., and Poole, F. G., 1975, Extension of the Cordilleran miogeoclinal belt to the San Andreas fault: Geological Society of America Bulletin, v. 86, p. 205–212.

Thompson, G. A., 1960, Problems of late Cenozoic structure of the Basin Ranges: Proceedings, 21st International Geologic Congress, Copenhagen, v. 18, p. 62–68.

Tyler, D. L., 1979, The Cordilleran miogeosyncline and Sevier(?) orogeny in southern California, *in* Newman, G. W., and Goode, H. D., eds., 1979 Basin and Range Symposium: Rocky Mountain Association of Geologists and Utah Geological Association, p. 75–80.

Wernicke, B., 1981, Low-angle normal faults in the Basin and Range province: Nature, v. 291, p. 645–648.

Wernicke, B., and Burchfiel, B. C., 1982, Modes of extensional tectonics: Journal of Structural Geology, v. 4, p. 105–115.

Williamson, D. A., 1980, The geology of a portion of the eastern Cady Mountains, Mojave Desert, California [M.S. thesis]: Riverside, University of California, 148 p.

Wise, W., 1969, Origin of basaltic magmas in the Mojave Desert area, California: Contributions to Mineralogy and Petrology, v. 17, p. 53–64.

Woodburne, M., and Tedford, R., 1982, Litho- and biostratigraphy of the Barstow Formation, Mojave Desert, California, *in* Cooper, J. D., ed., Geologic excursions in the California Desert (Geological Society of America Cordilleran Section meeting guidebook): Anaheim, California, Geological Society of America Cordilleran Section, p. 65–76.

Woodburne, M., Tedford, R., Stevens, M., and Taylor, B., 1974, Early Miocene mammalian faunas, Mojave Desert, California: Journal of Paleontology, v. 48, p. 6–26.

Woodburne, M., Miller, S. T., and Tedford, R., 1982, Stratigraphy and geochronology of Miocene strata in the central Mojave Desert, California, *in* Cooper, J. D., ed., Geological excursions in the California desert (Geological Society of America Cordilleran Section meeting guidebook): Anaheim, California, Geological Society of America Cordilleran Section, p. 47–64.

MANUSCRIPT ACCEPTED BY THE SOCIETY MARCH 4, 1986

ACKNOWLEDGMENTS

This work was supported by National Science Foundation Grant EAR-810752A. I thank B. Cox, K. Howard, J. K. Otton, R. H. Pilger, Jr., and an unknown referee for their careful and constructive reviews. Discussions with R. E. Anderson, G. A. Davis, E. G. Frost, R. H. Merriam, and M. O. Woodburne were invaluable. I also thank V. Iliff, C. Johnson, M. Kolb, M. Mahaffie, R. Mathis, and the late J. Meystayer for their able assistance in the field and in the laboratory.

Printed in U.S.A.

Processes of regional Tertiary extension in the western Cordillera: Insights from the metamorphic core complexes

William A. Rehrig
Applied Geologic Studies, Inc.,
2875 West Oxford #3,
Englewood, Colorado 80110

ABSTRACT

In the western United States, a major Tertiary extensional orogeny is distinguished from the more recent and subdued Basin and Range disturbance. This orogeny, characterized by dynamic horizontal translations (i.e., >100% extension in the Great Basin) occurred during the Eocene (~55–40 Ma) north of the Snake River Plain and during Oligocene–Miocene time (~35–16 Ma) farther south. Tectonic processes at shallow crustal levels included widespread listric or rotational faulting, decoupling along flat detachment faults, pervasive tensional fracturing, and calc-alkaline magmatism that yielded copious volumes of volcanic rocks and systematically oriented (north-northwest to north-northeast striking) dike swarms, veins, and elongate plutons. At deeper levels, the extensional deformation produced more dikes, intrusions, and the gently dipping mylonitic rocks exposed in Cordilleran metamorphic core complexes (MCC). Radiometric age criteria and ductile, kinematic strain indicators in the deeper rocks coincide with equivalent features in the brittly extended rocks above. The MCC are found in a regional setting of long antiformal axes of north to north-northwest trends, and transform discontinuities striking northeast to west-northwest, roughly parallel and normal, respectively, to the elongation of the MCC. Tertiary deformation in the complexes is commonly overprinted on earlier, gently dipping, compressional shear or metamorphic fabrics of Mesozoic to Paleocene age. Many MCC exhibit positive gravity anomalies and contain a preponderance of mafic dikes.

Deep-seated, mylonitic, normal fault zones have recently been cited to explain the Cordilleran MCC. This simple-shear explanation differs from the crustal stretching or boudinage model. Although attractive conceptually, the crustal shear-zone model in its present form has difficulty in sufficiently explaining described deformational fabrics and unique upper-plate lithologies restricted to the mylonitic complexes. Major displacements required of the model appear to be precluded by certain geometric constraints. Strain analysis has also been confused by superimposed compressional and extensional fabrics. New data from the Picacho MCC in southern Arizona support relatively shallow, in situ, Miocene mylonitization and detachment, and document the importance of pre-Oligocene low-angle deformation.

Various extensional mechanisms are proposed to explain flat detachment faulting. Mylonitic rocks exposed in MCC are derived from a setting of high heat flow and intrusion; preestablished, flat, crustal anisotropism and fluid-induced strain softening. Textural, isotopic, and geochemical evidence suggests that deuteric fluids were locally derived from the lower plate as a result of intense intergranular strain. These fluids, which concentrated or ponded at the detachment interface, may have enhanced upward mylonite development and are believed to be the principal cause for hydrothermal, chloritic brecciation overlying the mylonites. The MCC were uplifted by isostatic response to upper-plate denudation, lower-plate attenuation, and magmatic upwelling from below.

In terms of regional or plate tectonic setting, the extreme extension of the Tertiary orogeny is attributed to the incursion of hot asthenosphere into the Cordilleran crust above a segmented and sinking subduction slab. During the preceding Laramide orogeny, this oceanic slab had been driven shallowly under the North American crust with essentially no intervening mantle wedge. After about 40 Ma, dehydration fluids from the descending slab triggered magmatism throughout the Cordillera when the lower crust was contacted by hot asthenosphere. Mantle diapirism, convecting upward and laterally, became the fundamental mechanism for crustal softening and extension. The metamorphic core complexes may therefore represent local sites where mafic subcrustal material penetrated highest in the crust, causing the most visible effects of attenuation. As such, the MCC can be visualized as small-scale analogs of the extended Cordillera. It is likely that flat, stacked, en echelon mylonitic zones exist at deeper levels throughout much of the western United States.

INTRODUCTION

Only recently has the importance of a unique Tertiary extensional event become recognized by geoscientists working in the western Cordillera of North America. Parts of this tectonic history have been deciphered by an understanding of structures, metamorphic tectonites, and sedimentary features. Also, the greatly enhanced application of radiometric age dating to crystalline and metamorphic basement and faulted cover rocks has clarified many previously conflicting interpretations. These advances have gradually led to the distinction between a major Tertiary orogenic event and later Basin and Range tectonism.

In the central and southwestern part of the western United States, the major Tertiary extensional event has been called the mid-Tertiary orogeny by Damon and Bikerman (1964) and Rehrig (1981), and the Galiuro orogeny by Keith (1977); it has been reviewed recently by Elston (1984). South of the Snake River Plain the event spans from about 35 Ma (early Oligocene) to perhaps as late as 16 Ma (middle Miocene), depending on location. North of the Snake River Plain and extending into Canada, the extensional event is older, 40 to 50 Ma (middle Eocene). Its extensional and magmatic effects are slowly gaining recognition (Price and others, 1981; Harms and Price, 1983) amid the better-known backdrop of Laramide compressional tectonism. In the northwest, consequently, the extensional event has yet to be named.

To a large degree, the structures created during this Tertiary extensional deformation were formed within the brittle upper part of the lithosphere. They include: (1) "thin-skinned," listric fault blocks containing mildly to severely tilted or rotated rocks, (2) flat sole or detachment faults upon which bedding and listric faults merge or are truncated, and (3) transverse, transform-like zones searating domains of differing amounts of east-west extension. Widespread volcanism and epizonal plutonism accompanied the extension, and prominent dike swarms and elongate plutons document crustal dilation and spreading. The main distinction between this Tertiary extensional event and the younger, tectonically distinct Basin and Range extension is that the former resulted from a dynamic, horizontal pull-apart in a softened and hot lithosphere, whereas the latter was manifested by high-angle normal faulting in a relatively cold and rigid crust where vertical displacements exceeded horizontal movements and where magmatism was dominantly subcrustal.

To date, investigations have dealt largely with the brittle effects of the Tertiary extensional event, and for these there is growing documentation. Much less research and speculation have been devoted to deeper seated features or mechanisms responsible for the event. For example, the metamorphic core complexes (MCC) of the Cordillera, generally agreed to be expressions of ductile extensional deformation, were thought by some (Davis and Coney, 1979; Rehrig and Reynolds, 1980; Reynolds and Rehrig, 1980; Compton, 1980; Gans and Miller, 1983) to be expressions of ductile, Tertiary extension accommodated largely by pure shear. To others, the MCC were formed by predominant simple shear along major, low-angle, crustal fault zones (Wernicke, 1981; Davis and others, 1983; Davis, 1983). Critical problems that remain include (1) resolving the simple shear vs. pure shear argument, (2) determination of the true regional extent of the mylonitic zones and establishing their relation to surficial brittle deformation, and (3) definition of mechanical processes that form mylonites in the crust. This paper will examine these problems, considering the evidence at both shallow and deeper levels. In so doing, some newly recognized geologic and geochemical data from MCC will be integrated into the analysis, resulting in a model for the origin of the mylonitic complexes. Lastly, the Cordilleran-wide framework of the extensional event will be examined, by placing the MCC into a plate tectonic setting.

GEOLOGIC FEATURES OF THE EXTENSIONAL EVENT

Shallow Levels

Regardless of location or specific age (i.e., north or south of Snake River Plain; Eocene or Oligocene–Miocene), Tertiary extensional structures are remarkably similar throughout the mid-North American Cordillera. Within the uppermost crust these features, the results of brittle deformation, include: (1) Normal and listric-normal faults exhibiting moderate to extreme rotation of hanging wall blocks, (2) gently dipping to flat detachment faults with upper plates that are broken, fragmented, and tilted on a myriad of listric or rotated structures, (3) extensional sedimentary basins developed by normal, listric-normal, and detachment faulting, that were rapidly filled with distinctive assemblages of alkaline mafic volcanics, slope breccias, and coarse continental detritus, and (4) pervasive extension joints and close-spaced fractures in all types of rocks. In Tertiary plutons that were coeval with the Tertiary stress field, breakage is expressed as unidirectional extension fractures or fracture cleavage and, where filled by magmatic or hydrothermal products, as dike swarms, veins, or mineralized fractures.

The most diagnostic and influential structural effects (in terms of magnitude) of the extension are stratal rotations and tilted blocks associated with what can commonly be shown on geometric and observational grounds to be downward-flattening or curviplanar normal-fault surfaces. In many areas where this type of structure predominates, 100% or more crustal extension is indicated (Davis, 1983; Wernicke and Burchfiel, 1982; Gans and Miller, 1983; Miller and Gans, 1984). In the past these structures have been called denudational, dislocational, and listric-normal faults. Instrumental in their early recognition and description were Anderson (1971) in the southwestern United States; Armstrong (1972), Proffett (1977), and Wright and Troxel (1973) farther north in Nevada; and Longwell (1945) and Moore (1960) throughout the Basin and Range province.

In addition to curviplanar listric structures, Proffett (1977), Wernicke and Burchfiel (1982), and Gans and Miller (1983) emphasized rotational planar faulting as a more common mechanism for accommodating extension. In this process, planar faults rotate and flatten with increased tilting of strata, like a stack of tilted books in a bookcase. Geometric constraints and space problems inherent to either planar or listric rotation appear to seriously limit the extent to which beds can tilt and faults flatten without the generation of second- or third-order normal faults and compensation by penetrative brecciation. In fact, studies by Gans and Miller (1983), Colletta and Angelier (1982) and Gross and Hillmeyer (1982) described the myriad of small-scale structures that must accompany deformation of this type if extension is to approach 100%. To compensate for the open spaces which are geometrically required, major unconfined brecciation and low-angle fault splays (Gans and Miller, 1983) at shallow crustal levels become the dominant structures.

Detachment faults (occasionally termed decollements or dislocation surfaces) were originally described in strict coincidence with metamorphic core complexes (Crittenden and others, 1980). They were defined as major flat-lying discontinuities separating disrupted, tilted blocks of an upper plate from penetratively deformed chloritic breccia and ductilely deformed mylonitic rocks of a lower plate. They also separated domains of widespread thermal effects (K–Ar resetting) below from unperturbed rock above. The juxtaposition of lower-plate rocks exhibiting ductile, supposedly deep-seated, metamorphic textures against surficial Tertiary rocks showing no intergranular deformation has posed a major tectonic enigma which still lacks completely satisfactory explanation.

In detail, many detachment faults are not simple or single surfaces of displacement. Multiple, detachment-like faults are found well within lower-plate mylonites (Davis and others, 1982; Phillips, 1982). Slivers or lenses of brecciated, but generally nonmylonitic rocks bounded by detachment surfaces above and below occur in what might be visualized as medial positions. In the Whipple and Picacho MCC of California and Arizona, such nonmylonitic plates can be thought of as middle plates, because they occur sandwiched between upper-plate Tertiary supracrustal rocks and lower-plate mylonites. Middle plates were first described in the Buckskin Mountains of western Arizona (Wilkens and Heidrick, 1982). Some of these geometric complexities have been discussed by Lister (1984).

Detachment faults commonly bound upper-plate rocks, predominantly Tertiary volcanic and sedimentary strata, which display chaotic tilting, generally in one direction with consistently oriented rotational axes. Rarely, where detachment faults crop out on both sides of the arched mylonitic cores of some MCC, this tilt or tectonic transport direction is consistently maintained (i.e., Whipple MCC; Davis and others, 1980).

The distinction between (1) detachment faults and (2) gently dipping upper-plate faults should be examined, for the structures have been much confused. Recently, gently dipping fault surfaces have been described beneath variably tilted sequences of Tertiary supracrustal rocks at numerous localities in the Mojave and Sonoran deserts (Dokka, 1981; Garner and others, 1982; Berg and others, 1982; Mathis, 1982; Mueller and others, 1982; Lyle, 1982; Logan and Hirsch, 1982; Scarborough and Meader, 1983). These faults have been called "detachment faults" yet they commonly differ from those initially defined in MCC settings. In some cases these structures show moderately steep dips, and only slightly tilted upper-plate rocks (Garner and others, 1982; Berg and others, 1982). Usually there is the conspicuous absence of the detachment chloritic breccia zone below these faults, and mylonitic rocks are not present in the lower plates. The faults commonly exhibit rather limited lateral continuity. My own field work suggests that some of these structures have been misinterpreted and that a few are actually unconformities or upper-plate normal faults. In other cases, however, we are left with questions of definition for the flatter of these structures. Are they shallow or lateral equivalents of the MCC-related detachments at depth, as suggested by Wernicke (1981), or are they separate surficial features equivalent to the soles of listric faults or rotated planar faults?

Answers to these questions are difficult to document and depend somewhat on a model for the origin of detachment faults. The relationship most commonly described in the field and literature is that of multiple, listric or flattened normal faults that merge with or are truncated by the flat decollement or detachment. Upper-plate extensional structures initiated at higher angles have been progressively flattened (Gans and Miller, 1983), whereas detachment faults are assumed by most workers (for exception, see Davis, 1983) to have always been near horizontal. The normal and listric-normal structures are usually seen on top of, and generally juxtaposed with, the more extensive detachment discontinuity at depth. These observations suggest that the two kinds of structures are separate and distinct, and that the detachment is the lower decoupling surface upon which the upper plate breaks up or is transported. However, it should be pointed out that mylonitic detachment zones may grade laterally into nonmylonitic detachments. This was first suggested by Anderson and others (1983) and is documented in the Snake Range by Gans and others (1985) (see Fig. 1). In the Snake Range, the disappearance of ductile fabrics does not correspond to a cutting up-section (shallowing) of the detachment fault. This geometry departs significantly from that of the model of Wernicke (1985). Other possible mechanisms for detachment-like faulting are shown in Figure 1. As can be seen, they need not all be accompanied by mylonite development.

The brittle extensional effects discussed above are the direct or indirect results of strain focused along distinct zones or surfaces of tensile failure. The overall rock mass situated between shear surfaces shows the effects of a more homogenous, distributed style of brittle tensile strain. Tertiary (Eocene to Miocene) epizonal plutons are pervasively broken by vertical fractures, joints, and veins (Fig. 2B), which are oriented perpendicular to the regional extension direction (Rehrig and Heidrick, 1976). Emphasized in fine-grained, homogeneous rocks at near-surface levels, the tensile breakage commonly resembles a fracture cleavage or shattered type of jointing characterized by "open," short, curviplanar surfaces. Within Laramide plutons of Arizona, this type of high-level fracturing was recognized as clearly postdating the Laramide-age structures (Rehrig and Heidrick, 1972) (Fig. 2A). Not only did the post-Laramide, pervasive, tensile stresses thoroughly shatter these rocks, but the mass movement of rock in extension (east-northeast–west-southwest in Arizona) imparted low-angle, subhorizontal movement indicators (i.e., grooves, striations, slickensides) upon east-northeast to east-west veins and fracture surfaces (Fig. 2A) which had earlier formed as Laramide dilational or extensional structures (Rehrig and Heidrick, 1972, p. 211).

On a regional scale, Tertiary plutons and dike swarms reflect the extensional deformation (Fig. 2C). Although orientations of these features vary locally due to preexisting anisotropic weaknesses or stress deflections, the overall regional strike of these dilational structures parallels that of the smaller scale fractures or cleavage, and the strike of tilted strata. The direction of extension so defined is roughly east-northeast for the southwest; east-west to west-northwest for much of the middle Cordillera, and largely west-northwest for Eocene extension in the northwest (Rehrig and Heidrick, 1976; Zoback and others, 1981; Gans and Miller, 1983; Rehrig and Reynolds, 1981.)

Deeper Levels

We are far less certain of the effects of Tertiary extensional tectonism beneath the shallow lithosphere. The reconstituted, mylonitic, basement fabrics exposed in the MCC of the North American Cordillera may represent the result of ductile normal shear or stretching at depth—an analog to the brittle extension described above. There is little doubt that original protolith textures, and structures, have been largely transposed and remade into low-angle, foliated and lineated mylonitic gneisses (Crittenden and others, 1980), but the timing, stress systems, and mechanisms for these fabrics are much less definite. These uncertainties, coupled with the persistent evidence for Mesozoic deformations in the MCC, have formed the basis for continued debate on age, depth, and fundamental origin of the mylonitic terranes.

As reviewed by Rehrig (1982), it appears certain that compressional, shear-induced mylonites of Laramide age crop out in the Cordillera; in fact, they seem to be retained within MCC exposures. At this point, however, unequivocal evidence for Tertiary extensional mylonitization should be cited. In the northwestern United States, mylonitic fabrics clearly affect Eocene plutons. This has been documented for the Swimkin Creek pluton in the Okanogan complex (K. Fox, 1983, personal commun.) and for the Silver Point stock in the Priest River (Spokane) MCC (Harms, 1982). The Silver Point body underlies the Newport detachment fault. Similar mylonitic textures overprint a variety of Precambrian, Cretaceous, and probably Eocene intrusions along the eastern contact of the Priest River MCC with the Purcell trench. These gently east-dipping mylonites appear to increase in thickness southward toward the Lewis and Clark lineament in the Coeur d'Alene Lake area (Rehrig and others, 1986b). South of the Lewis and Clark zone, the Bitterroot "front" is a 1.5-km-thick, 100-km-long mylonite zone within plutonic rocks of the Idaho batholith. Until recently, the affected plutonic terrane was considered to be of Late Cretaceous–Paleocene age (Hyndman, 1980). New U-Pb zircon and Ar-Ar hornblende dating of the mylonitic crystalline rocks now indicates that the terrane is made up largely of 40–50 Ma mesozonal to epizonal intrusions which shortly predate or are synchronous with the pervasive ductile deformation (Chase and others, 1983; Garmezy, 1984).

In the Grouse Creek and Raft River MCC of northwestern Utah, Eocene to Miocene ductile flattening and east-west extension are documented within plutons as young as 24 Ma (Todd, 1980; Compton, 1980; Compton and others, 1977). Gans and Miller (1983) documented a similar history in the Snake Range

A. Intrusive Generated

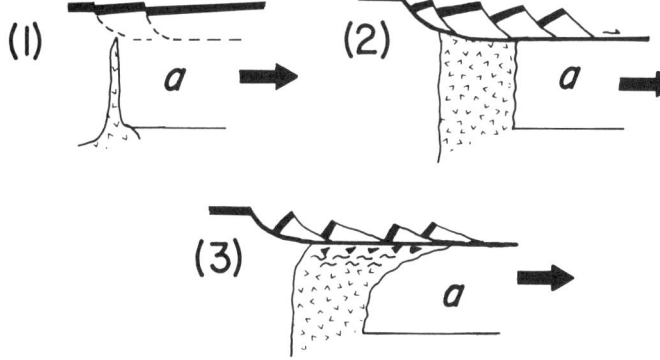

B. Ductile Normal Shear Zone

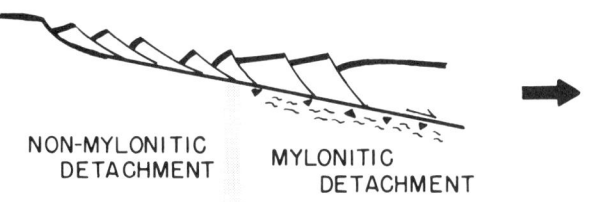

C. Crustal Stretching

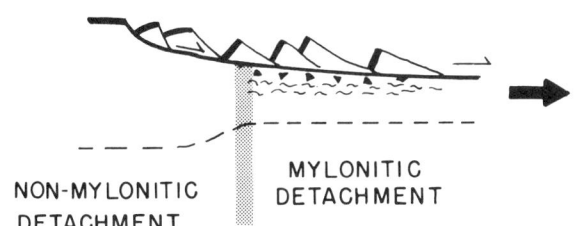

D. Middle Plates

E. High-level Listric Sole Fault

F. Explanation

░░	INTRUSIVE BODY
▲▼▼	CHLORITIC BRECCIA
━━↙	DETACHMENT FAULT
～～	DUCTILE MYLONITE FABRIC
➡	DYNAMIC TENSIONAL STRESS FIELD

Figure 1. Mechanisms for generating low-angle detachment faults. In A, (1) and (2), detachment is formed by dikelike pluton emplacement. Near-surface block *a* moves to right by dilation from intrusive and regional extension. Flat shear surface is created by this movement between dilated block and surficial layer above the intrusion. In (3), mylonitic fabrics and further detachment may be caused by ductile stretching of still softened, water-rich upper part of pluton or its roof. In B, detachment fault is equivalent to ductile, normal fault zone as proposed by Wenicke (1981). In C, detachment is a result of intracrustal stretching in thermally softened rocks modeled after explanations by Rehrig and Reynolds (1980) and Gans and others (1985). Note that ductile fabrics along detachment give way laterally to a brittle shear surface without cutting up-section toward the surface. In D, a nonmylonitic detachment is formed by a relatively thick middle plate. Mylonites are not exposed but exist at depth below a more fundamental detachment zone. In E, a detachment-like structure is merely the downdip, flattened part of a high-level, listric normal fault. As such, it is entirely an upper-plate or surficial structure of relatively local extent and has little direct relationship with metamorphic core complex features. In F, symbols used in A–E are explained.

of eastern Nevada. In Arizona, Rehrig (1982) has reviewed evidence for post-25 Ma lineated mylonitic textures imposed on plutons of approximately that age in the South Mountain, Picacho, and Catalina-Tortolita MCC. In addition, post-50 Ma mylonitization is indicated in the Harcuvar complex of western Arizona. Dating of mylonitized plutons as young as 50 Ma in Mexico supports Tertiary ductile strain in that province (Anderson and others, 1980).

Considerable study and documentation (Davis and others, 1975; Davis, 1975; Compton, 1980; Rowles, 1982), show that these Tertiary mylonites are the products of varying proportions of flattening and approximately east–west extension and shear. The extension is further evidenced by the presence and orientation of roughly consanguinous vein and dike patterns striking perpendicular to the trend of ductile elongation and lineation in the mylonites (Fig. 3). These dilational patterns occur within and outside the mylonitic domains and establish beyond doubt that the lineated, foliated fabrics formed in a regionally consistent and synkinematic extensional regime (Fig. 2).

A number of peculiarities of Tertiary mylonitic terranes have come to light recently, which relate to the forthcoming discussion of extensional processes. They are discussed briefly below.

At both local and regional scales, mylonite development

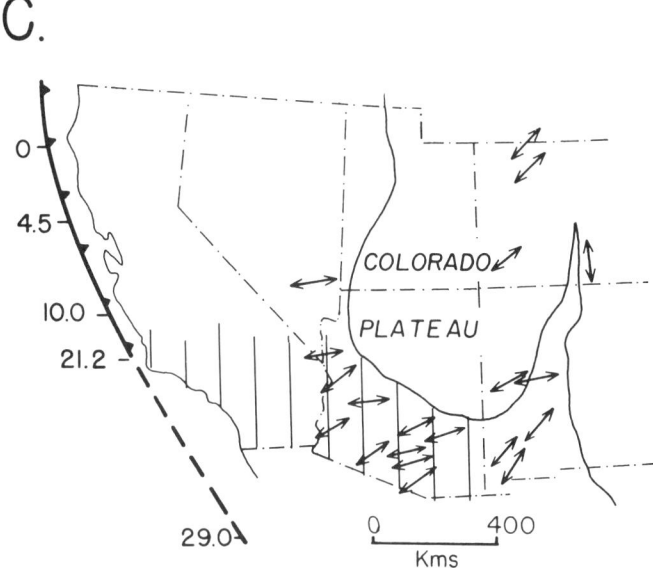

Figure 2. (This and previous page) A: Photo showing distributed extensional strain recorded in granitic rock of southern Arizona. The straight continuous joints trending toward the observer are of Laramide age and are filled with hydrothermal products from the consanguinous pluton. In contrast, the irregular, curviplanar, discontinuous cross features are decidedly later, and are probably a result of high-level crustal extension. This extensional fracturing of regional, north-northwest-strike affects all rock types and appears most intense in fine-grained to aphanitic rocks. Slickensiding and striations on east-northeast to east-west-striking, Laramide extension joints are subhorizontal and are believed to be a product of Tertiary extension and regional east-west expansion of rock mass. B: Pervasive north-northwest-trending, extensional fracture fabric etched in a middle Tertiary (Miocene) pluton of southern Arizona, as shown from high-altitude air photo. C: Regional extension directions through the Southwest cordilleran crust during a portion of the middle Tertiary extensional event (25 to 10 Ma). Data taken from Zoback and others, 1981. Note that the least principal stress vectors, as defined by orthogonal dike swarms, are roughly parallel to lineation in the Tertiary metamorphic core complexes, even though the dikes recorded were not within the complexes. Numbers along subduction zone show the southern extent of subduction (megayears). Region vertically lined was released from subduction between 30 and 15 Ma.

seems to be enhanced along preexistent gently dipping to horizontal lithologic or structural discontinuities. At small scale, flat quartz veins, dikes, or lithologic contacts (i.e., flat, tabular xenoliths of schist in granitic host rock) appear to promote the lineated foliate Tertiary fabrics. At larger scale, some Tertiary mylonite zones occur above or in the uppermost parts of Tertiary, high-level plutons and below their extrusive equivalents. There is increasing realization that Tertiary mylonitic deformation commonly is superimposed on low-angle fabrics of Mesozoic age. Relatively few MCC have escaped this coincidence, and it represents the primary cause for much past confusion and debate concerning age and origin of these tectonic features (i.e., Davis and others, 1980; DeWitt, 1980; Mattauer and others, 1983; Brown and Read, 1983). In metamorphic core complexes such as the Shuswap, Priest River, Albion–Raft River–Grouse Creek, Snake, Ruby, Whipple(?), Harcuvar, and Picacho, there is fairly definitive evidence that middle Mesozoic to Laramide orogenesis involving relatively deep-seated magmatism (2-mica) and metamorphism, thrusting, ductile folding, and foliation transposition converted regional basement trends of high-angle structural fabric to domains of low-angle anisotropism (i.e., Okulitch, 1984; Fox and others, 1976; Rehrig and others, 1986b; Davis and others, 1982; Rehrig and Reynolds, 1980; Rhodes and Hyndman, 1984). The creation of these areas of subhorizontal fabric appears to have "prepared the ground" for later Tertiary extension and, under suitable conditions of strain rate, temperature, pressure, etc., superimposed extensional mylonitization was preferentially developed (Rehrig, 1982) (Fig. 4). This coincidence of Mesozoic and Tertiary low-angle tectonic fabrics was the subject of recent analysis by Coney and Harms (1984).

At the time of the Geological Society of America memoir on metamorphic core complexes (Crittenden and others, 1980) most workers, impressed with the striking structural and metamorphic contrasts between upper and lower plates, favored a deep-seated (mid-crustal) and pre-Tertiary setting for lower-plate deformation with respect to detachment and upper-plate extension. Therefore, juxtaposition of these two highly incongruous regimes implied great vertical displacements of rock mass, realistically only achievable through a fairly extended period of geologic time. Subsequently, because of radiometric dating of relatively high-level, Tertiary plutonic protoliths in lower-plate mylonite zones (Harms, 1982; Chase and others, 1983; Rehrig, 1982), both the depth of mylonite formation and its separation (time and space) from detachment events have dramatically decreased. A direct and genetic association between ductile mylonitization and brittle detachment is implicit in the most recent models for the MCC (Davis, 1983; Davis and others, 1983; Davis and others, 1986).

Through many areas of the extended Cordillera, pre-Tertiary rocks lying beneath the level of surface volcanism are intruded by abundant dike swarms and elongate intrusive bodies of Tertiary age. The intrusions appear to be more numerous in basement rocks than in volcanic cover; however, this may be a matter of ease of recognition. As shown by Rehrig and Heidrick (1976) and Zoback and others (1981), these dilational features are systematically oriented perpendicular to regional σ_3 (least principal) stress.

Many MCC are sites of particularly intense dike emplacement oriented normal to extensional lineation (Fig. 3B). In places (i.e., Whipple and Harcuvar MCC), the dikes concentrate at the west end of the complexes. A common dike type consists of microdiorite of fairly primitive composition and strontium isotopic ratio (<0.706; Rehrig and Reynolds, 1980). The dikes have been dated at 27 Ma (hornblende) and 22 Ma (biotite) in Arizona where they postdate mylonitic fabrics (Rehrig, 1982). I recently noted deformational fabrics in these bodies at the very top of the mylonite sequence in the easternmost Harcuvar MCC of western Arizona.

The passive or dilational emplacement of magmas into an extending crust is therefore a deeper level expression of the Ter-

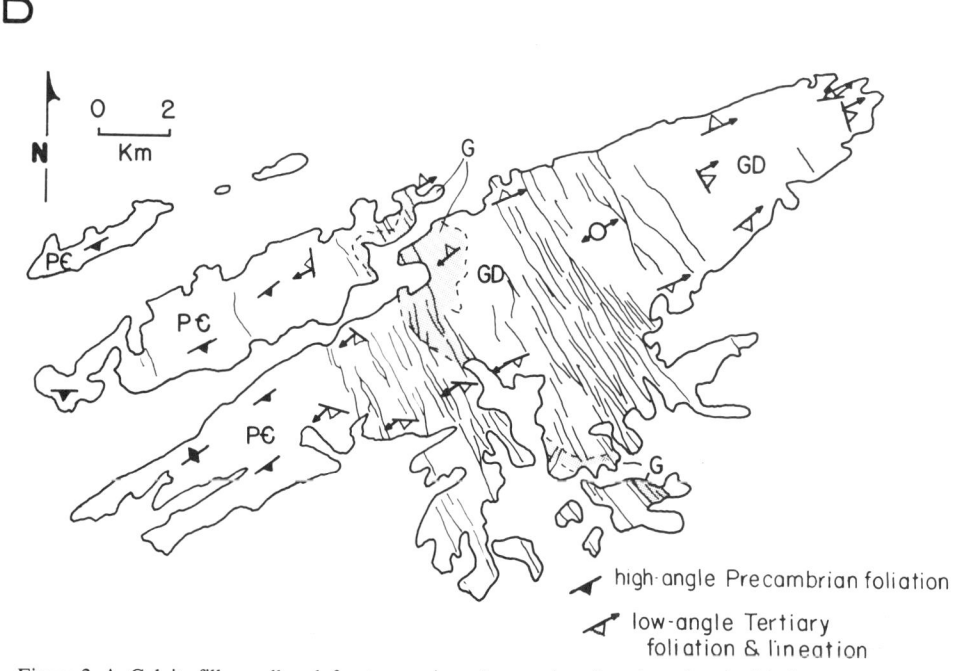

Figure 3. A: Calcite fills small gash fractures oriented normal to direction of mylonitic lineation in hand specimen from Tertiary metamorphic core complex in Arizona. B: Dike swarms (20–25 Ma) oriented normal to foliation arch and mylonitic lineation in the South Mountains metamorphic core complex, south of Phoenix, Arizona. Dikes and mylonitic fabrics cut the 25 Ma granite (G), granodiorite (GD), and Precambrian metamorphic and igneous rocks (P€). Some dikes become mylonitic at uppermost levels of exposure. Data from Reynolds, 1982.

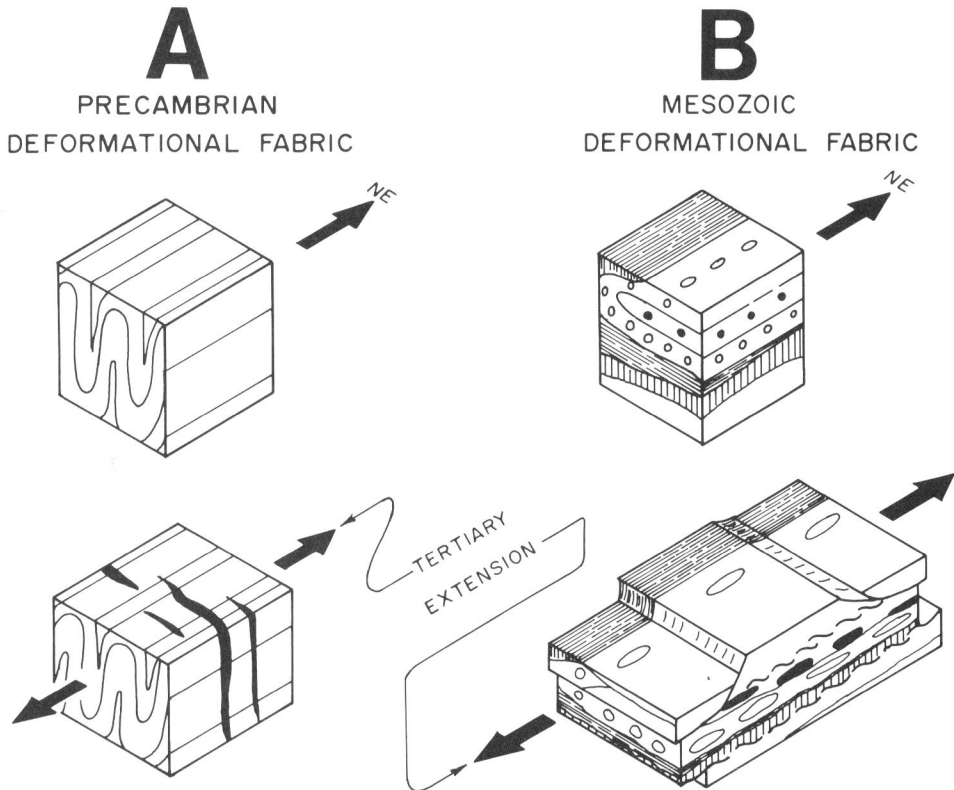

Figure 4. Diagrammatic illustration of the effects of preexisting low-angle, anisotropic fabric on Tertiary extension. In A, the rocks retain their northeast-trending, vertically isoclinal, Precambrian corrugation, inhibiting the development of flat Tertiary structures. In B, because of earlier, Mesozoic Laramide low-angle structural and metamorphic effects, Tertiary extension is enhanced.

tiary extension. The dikes and plutons are sufficiently numerous in certain domains that they may cumulatively represent significant amounts of extension. Note from Figure 1A the influence that pluton emplacement may have on the initiation of subhorizontal detachment faults. Obviously, the importance of this extensional mechanism depends on the size of the area being considered and the spacing and frequency of the dikes, which is highly variable from place to place. My experience, based on numerous field studies, is that the extent of plutonic dilation is usually underestimated. This is primarily because we have not mapped in sufficient detail or over enough areas.

REGIONAL FEATURES OF IMPORTANCE

Regional analysis of the attitudes of tilted Tertiary supracrustal rocks through the Great Basin and Sonoran provinces of the Cordillera reveals domains in which direction of tilt is consistent from mountain range to range. Furthermore, these domains define large-scale antiforms and synforms which can be traced for hundreds of miles along their axial strike (Rehrig and Heidrick, 1976; Stewart, 1979). However, these unusual features have no real structural relief so they cannot be construed as anticlinal or synclinal structures.

The regional antiforms wrap around and conform to the margin of the Colorado Plateau (refer to map in Stewart, 1979). Their axes are also roughly parallel to elongation of the MCC, and thus they form tectonic trends which are roughly perpendicular to lineation or the extension direction in the complexes. At several places in Arizona, the antiforms lie along or correspond to the MCC arches (Rehrig and Heidrick, 1976). Elsewhere, this correspondence is not maintained, and the MCC appear to interrupt a surrounding domain of consistent, homoclinal-like tilt. Thus, a paradox of sorts is formed by some tilt domains possessing a clear spatial association with mylonitic zones of MCC, whereas through other areas of the Great Basin, antiformal structures show no obvious relationship with mylonitic terranes, at least as can be ascertained from surface exposures.

The antiformal axes and tilt domains appear sharply truncated along ill-defined boundaries of approximately east–west to northeast strike (Stewart, 1979). These discontinuities were first noted in the Rio Grande rift (Chapin and others, 1978), where they divide areas of mutually opposing east- and west-dipping Tertiary strata. They have also been described in Arizona by Rehrig and others (1980) and through the greater Basin and Range province by Stewart (1979). These features seem as indigenous to the extended Cordillera as the regional antiforms, and at several places the discontinuities form transverse boundaries to metamorphic core complexes. An analog to these discontinuities in Idaho and Montana is a segment of the Lewis and Clark lineament, which forms a line roughly connecting the Priest River (Spokane dome) and Bitterroot MCC (Rehrig and Reynolds, 1981). Efforts to equate many of these discontinuities with traces of discrete faults have thus far failed. Therefore, the transverse boundaries can only be defined as diffuse zones or ill-defined boundaries separating areas of opposite structural tilt direction. It should be noted that the trends of the boundaries maintain regional perpendicularity with respect to the antiformal axes and trends of MCC.

The interrelationship between antiform, MCC, and transverse discontinuity has been described in western Arizona, where it is particularly well illustrated (Rehrig and others, 1980). This example suggests that the transverse discontinuity, which here offsets the antiform, is a transform boundary separating regimes of differential extension (Fig. 5). This interpretation might apply equally to the Lewis and Clark zone, but it is not as well documented elsewhere.

ANALYSIS OF THE DUCTILE SHEAR ZONE MODEL

Ramsay (1980) and Wernicke (1981) have proposed that ductile shear zones of shallow dip are a reasonable means to accommodate crustal extension at both shallow and deeper crustal levels. Subsequently, workers (Davis, 1983; Davis and others, 1983; Davis and others, 1986) have suggested a model equating these shear zones with the Cordilleran metamorphic core complexes. The original model of Wernicke was put forward merely as a working hypothesis. However, with the relatively recent recognition of S-C (shear and flattening) fabrics in the mylonitic rocks beneath detachment faults (Lister and Davis, 1983), there has been a movement by many workers toward the uniaxial simple shear zone model. According to this model, the mylonites of MCC are deep-seated, ductile shear zone rocks that have been juxtaposed with surficial upper-plate rocks by substantial, low-angle normal-fault displacement.

As reviewed by Wernicke (1985), the shear zone model is attractive because it appears to offer explanations for relationships problematic to models based largely on crustal stretching. Chief among these problems is the apparent age discrepancy between blocking temperature dates in the lower-plate tectonites and the youngest ages of detachment activity. In the southwestern United States, this timing difference can amount to as much as 9 10 m.y. (approximately 25 to 16 Ma). This difference, however, may merely be an artifact of a time and temperature deformational variation through the deformed interval.

Blocking temperature ($\sim$250°C) dates on biotite in the mid-20-Ma range are recorded from the mylonites. As pointed out by Rehrig (1982), however, these apparent ages decrease significantly near the actual detachment fault where they closely approximate the age of the youngest rocks tilted upon the fault ($\sim$18 to 16 Ma). It thus would seem that although rocks well below the detachment surface cooled below 250°C perhaps as early as 25 Ma, ductile deformation and thermal effects above continued well into the period of brittle detachment. Perhaps a question equally difficult to the shear zone concept is why thermal resetting is so pervasive through hundreds and even thousands of meters of weakly mylonitic to nonmylonitic rocks well below the detachment faults in some MCC (i.e., Miller and Engels, 1975; Davis and others, 1982), or in the case of the Snake Range MCC, why neither thermal nor deformational effects seem to significantly diminish with depth into the lower plate (P. B. Ganz and E. L. Miller, 1985, personal commun.).

A further problem inherent to a stretching concept for MCC are "detachment faults" not underlain by ductile mylonites. Actually, the recognition of these features may have little direct bearing on the mylonitic complexes themselves. As dicussed earlier, these flat faults may be distinct from MCC features and therefore may be of somewhat different origin (See Fig. 1).

Fabric and strain analysis within the lower-plate mylonites is still in its infancy. Preliminary studies have been prone to misinterpretation because of mixed multiple deformations, and results so far have commonly been inconsistent or contradictory. Earlier work cited orthorhombic strain symmetries suggestive of stretching or pure strain (Compton, 1980; Snoke, 1980). Rankin (1980) emphasized this style of strain, especially from the lower parts of the mylonite zones, and Gans and Miller (1983) cited orthorhombic strain as characterizing the Snake Range MCC. In contrast, Gzrmezy (1984) documented large amounts of monoclinic strain in the Bitterroot mylonite front. As mentioned, S-C structures recently recognized from the deformed zones also attest to uniaxial simple shear. The initial ductile shear zone model of Ramsay (1980) included secondary flattening (s-surfaces), and a model based on nonuniform, layer-parallel, pure shear should probably generate common secondary shear structures (Fig. 6). Realistically, therefore, it would not be surprising if strain analysis would document both pure and simple shear regardless of the predominant process. Exceptionally rigorous and detailed work is required to determine which mechanism is the principal one.

Furthermore, confusion has resulted because of the difficulty in recognizing and separating the effects of Mesozoic ductile deformations which coexist with those of Tertiary origin. Although the rock strains originated from highly distinct tectonic episodes (i.e., compressional, extensional), the resultant fabrics are nearly indistinguishable. Particularly confusing is where Tertiary extensional mylonitization has been localized along thick zones of ductile thrusting marked by well-developed S-C mylonites (Haxel and Grubensky, 1984). Structural superposition of this type obviously could lead to inappropriate structural conclusions unless age constraints on the fabrics are first established. Other problems have developed where crystalloblastic (metamorphic) and mylonitic fabrics have not been distinguished. Examples where such difficulties exist include the Okanogan and Priest River MCC in Washington (Fox and others, 1976; Rhodes and Hyndman, 1984; Rehrig and others, 1986B), the Whipple and Harcuvar complexes in the southwest (cf. Davis and others, 1980; Rehrig and Reynolds, 1980) and the Picacho MCC in southern Arizona (S. B. Keith and Rehrig, unpub. data).

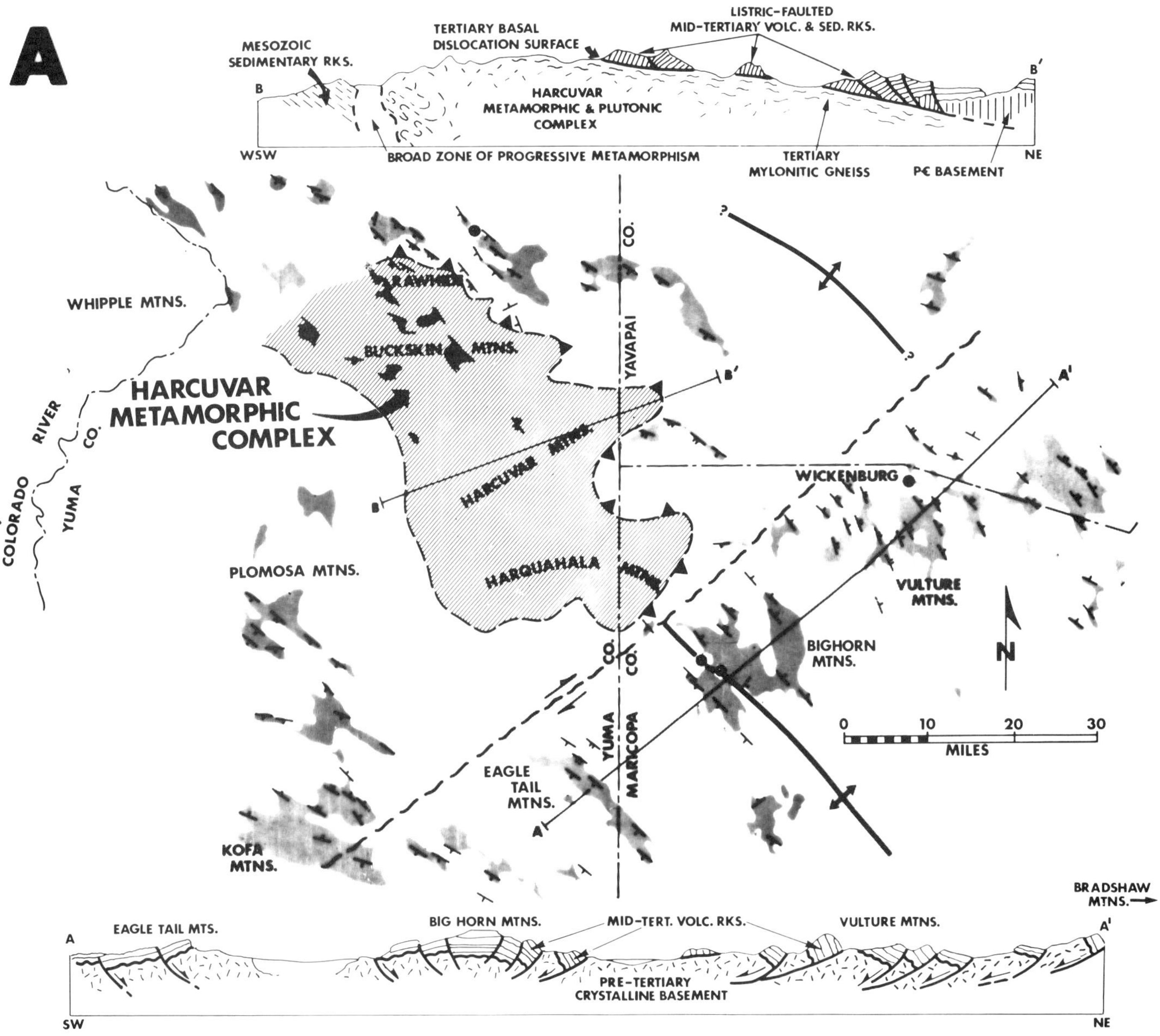

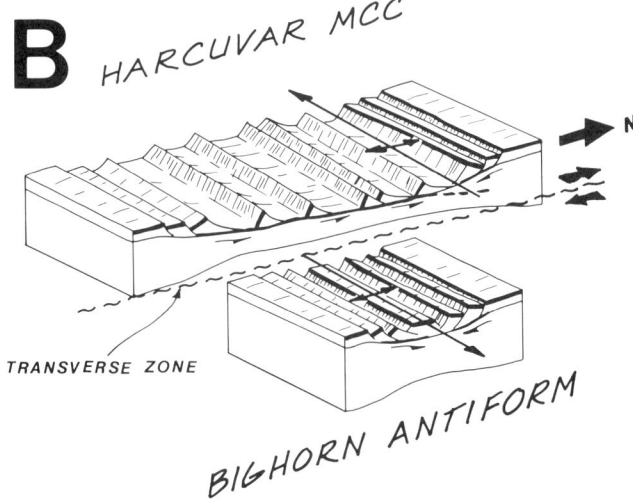

Figure 5. Possible relationships between metamorphic core complex (MCC), regional antiforms, and transverse discontinuity (transform) in west-central Arizona. A: Tectonic sketch map of Harcuvar MCC and surroundings showing areas of tilted Tertiary supracrustal rocks, antiform axes, and ill-defined transverse zone separating domain of opposing stratal tilt. Cross sections across Bighorn antiform and Harcuvar MCC are schematic. Data from Rehrig and others, 1980. B: Schematic isometric diagram illustrating transverse discontinuity depicted as a transform-like boundary between domains of differential crustal extension. The Harcuvar domain stretched or extended much farther northeast than did the Bighorn block.

A. Low-Angle Ductile Shear Zone

B. Non-Uniform Ductile Flow (stretching)

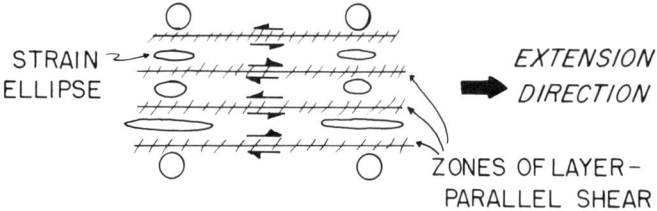

Figure 6. Comparison of mixed secondary simple and pure shear deformations by primary mechanisms of ductile, normal shear zone or layer-parallel, nonuniform ductile flow. Note that in either case, coexistent simple/shear and flattening fabrics may be expected.

To base a tectonic model for MCC and Cordilleran extension *solely* on either simple or on pure shear is a temptation which should be avoided. In a tectonic setting characterized by widespread magmatism, high heat flow, crustal softening or thinning, and dynamic regional extension, coeval zones of ductile stretch *and* simple shear may develop.

In the case of the MCC, a number of characteristics are not readily explained by a unique simple shear model. They are described as follows:

1. Zones of Dominant Flattening and Coaxial Pure Shear

Perhaps the most spectacular documentation of nonuniform flattening and extensional structures associated with mylonitic lower-plate rocks is described by Davis and others (1982) in the Whipple MCC of southeast California. Nonsheared deformational fabrics in dikes within and above the west-dipping mylonitic front (refer to their Figures 6A–6C) "record a protracted history of coaxial strain." Not far below the mylonitic front, steeply dipping Precambrian compositional foliation is shown in detail to be overprinted by recumbent buckling and layer-parallel shortening which accompanied development of middle Tertiary extensional lineation and foliation. The authors emphasize the lack of simple shear. This writer has noted such flattened folds within Pinal Schist beneath the Guild Wash detachment fault at the north end of the Tortolita Mountains near Tucson, Arizona. In like manner, low-angle, ductile normal faults described by Davis and others (1982, Fig. 7) occur in a medium of pinch-and-swell and boudinage structures indicative of brittle-ductile variations within a setting of flattening and northeast-southwest extension. G. H. Davis (1980, p. 53–55) also discussed and meticulously illustrated these kinds of structures (his Fig.

9A). Davis and others (1975) and Davis (1980) carefully evaluated the nature of strain in the mylonitic carapace of the Tortolita Mountains in southern Arizona, comparing deformation fabrics in granitic rocks with fold and stretched-pebble fabrics in late Precambrian conglomerates. They concluded that deformation was coaxial pure strain with subhorizontal flattening and northeast–southwest extension of rock mass in axial ratios of 9:2:1.

An impressive argument for co-axial extension (i.e., nonuniform pure shear) results from study of the Newport detachment fault in northeast Washington. This spoon-shaped, flat structure with a horseshoe-shaped trace overlies mylonitically deformed Precambrian and Eocene crystalline rocks of the Priest River MCC (Miller, 1971). A rigorous structural analysis by Harms (1982) has established that the fault has moved in a low-angle normal sense along both the east and west branches of the structure (Fig. 7). The remarkable aspect of this movement picture is that the sense of normal slip of the upper plate, as measured along east and west sides, is convergent (Fig. 7). This constitutes a geometric and mechanical improbability if the upper plate is the active block. If, however, the active motion is concentrated in the ductile lower plate, then the sense of motion must be coaxial stretching in opposing directions and cannot be construed as a case for overall uniaxial simple shear (Fig. 7).

It is of interest that a number of workers (some cited above), having so elaborately documented the abundant evidence for coaxial stretching and flattening, are now among those who, after the recognition of S-C shear fabrics, strongly advocate the simple shear model for MCC.

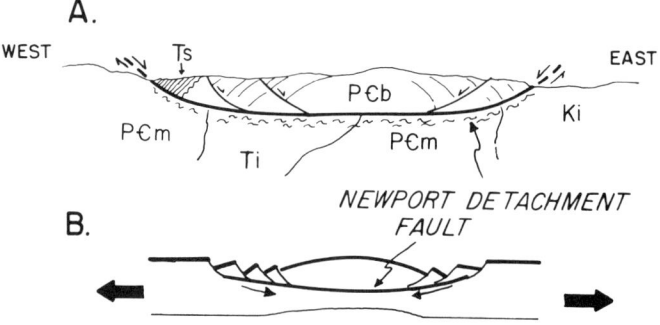

Figure 7. Evidence for coaxial crustal stretching beneath the Newport detachment fault, Priest River MCC, northeast Washington–northern Idaho. A: Schematic cross section through the southern part of the Newport allochthon. P€m = lower plate metamorphic and plutonic rocks; Ki = Cretaceous Kiniksu batholith; Ti = Eocene Silver Point pluton; P€b = Proterozoic Belt sedimentary rocks; Ts = Eocene continental sediments. Data for normal fault motion at east and west ends of Newport structure are from Harms (1982). Note that motion arrows on the lower plate are only relative to normal-fault indicators of upper plate. B: Diagrammatic model for development of the Newport detachment by large-scale crustal boudinage (Price and others, 1981; Harms, 1982). Arrows in ductilely thinned lower plate represent absolute motion of material which transports listric faulting within overlying brittle plate.

2. Foliation Attitudes Through the Mylonite Zone

Field examples and theoretical aspects of ductile shear zones cited in Ramsay (1980) and Ramsay and Graham (1970), indicate a geometric pattern of progressively steeper dipping foliation with depth below the shear. This would require that lower, weaker developed mylonitic fabrics in the MCC have steeper dips than those vertically above at the top of the zones. This relationship is not generally documented in MCC where much of the mylonitic zonation is exposed. Although the overall dip asymmetry noted across the entire length of many MCC foliation arches (see Crittenden and others, 1980) could be cited as evidence for this shear configuration, it is a horizontally separated asymmetry, not one noted through a vertical column at any one position along the mylonitic zone. In two cases (i.e., Harcuvar and Priest River MCC) where such lateral asymmetry is well established, the steeper foliation is of a metamorphic or crystalline nature, and has a distinct direction of lineation. This fabric is thought to be decidedly older than the discordantly superimposed, more mylonitic lineation and foliation which is developed above and to one side (Rhodes and Hyndman, 1984; Rehrig and Reynolds, 1980; Rehrig and others, 1986B).

3. Lateral Continuity

If the mylonite and detachment zones of the MCC are representative of low-angle, crustal, normal faults on the scale proposed by Wernicke (1981), their two-dimensional continuity should be much greater than can, in some cases, be documented from regional mapping. In some MCC it is difficult or impossible to trace detachment faults even around a single range. In certain examples, well-developed detachment zones appear to "play out" or disappear over short distances (1–3 km) along strike. Examples include the Newport fault at its north end (Harms, 1982) and the Purcell detachment zone just south of Sandpoint, Idaho (Rehrig and others, 1986B). The MCC appear to be relatively isolated occurrences surrounded by ranges where basement rocks exhibit little evidence of mylonites, detachment faults, or seriously tilted rocks. Examples of these sudden lateral discontinuities in Arizona and southeast California have been cited by Rehrig (1982) and are also documented by Gans and others (1985) in eastern Nevada.

4. Relatively Unsheared Dikes

This problem for simple shear is an outgrowth of point 1 above, but it focuses on the observation that synkinematic subvertical dikes oriented normal to extensional fabrics in lower plates of MCC are commonly not seriously sheared as should be expected in a major simple shear zone (Fig. 8A). Relationships such as are shown in Figure 8B are observed in the South Mountains MCC (Reynolds, 1982), are described by Davis and others (1982), and have been noted by this worker in the Picacho MCC (see Fig. 10C). One might argue that dike contacts are undigi-

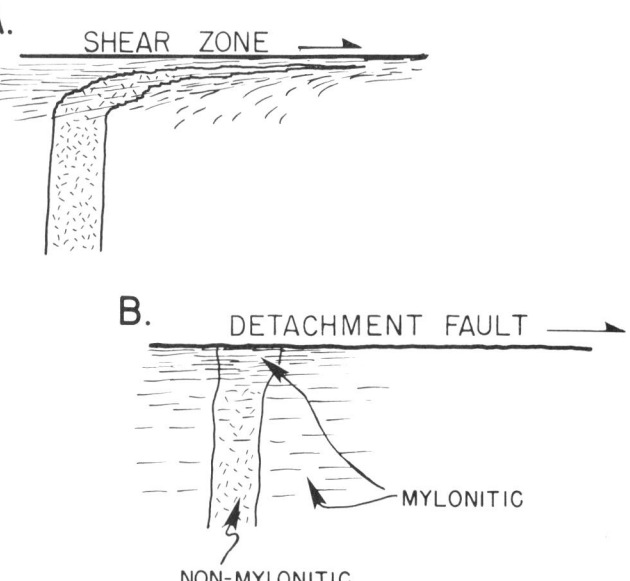

Figure 8. Comparison of dike contacts within mylonite zones. A: Sheared dike within a ductile shear zone, patterned after relationships cited by Ramsay and Graham (1970). B: Schematic representation of dike approaching the Tertiary detachment surface. Commonly dikes are undeformed lower in lower plate where they cut earlier developed Tertiary mylonitic fabrics in host rocks. Higher, the dikes become mylonitic, but contacts are not significantly sheared or disturbed.

tated or not seriously sheared because the dikes were intruded so late in the shearing time interval. However, many of these dikes are noted near the uppermost and latest parts of the mylonitic sequence, and dike textures display intensely mylonitic internal fabrics. It is thus quite difficult to imagine them escaping the deformation as depicted in Figure 8A, if the mylonite formed by predominant simple shear.

5. Geometric Constraints

The inability to substantiate the large lateral displacements or the kilometres of missing upper crustal rocks seriously hinders the shear zone model, at least at larger scales. These problems have been brought into focus recently in the Snake Range MCC by Gans and others (1985). Another example is in the Whipple-Rawhide area of southeast California and westernmost Arizona. A diagrammatic cross section from the Turtle Mountains just west of the Whipple MCC to the Colorado Plateau (Fig. 9) illustrates the geometric constraints. The detachment shear is thought to terminate in a breakaway zone in the area of the relatively untilted Turtle and Mopah mountains (Howard and others, 1982b). Along the entire known length of the detachment surface, a distance of about 100 km, Tertiary supracrustal rocks occur in upper-plate positions and have overall consistent tilts indicative of northeast transport. To displace these rocks in normal-fault fashion on a major low-angle crustal shear from surficial positions to juxtaposition against underlying mylonitic

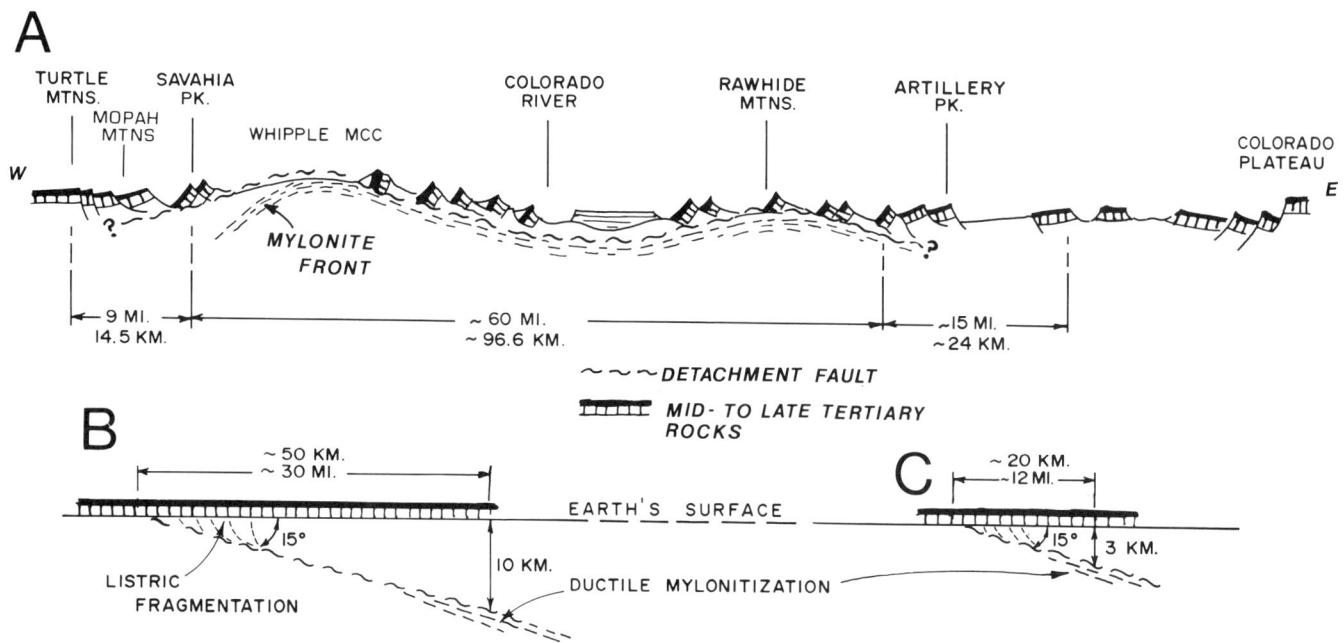

Figure 9. A: Schematic structure section from eastern California to the Colorado Plateau of Arizona, illustrating geometric constraints to mylonitic detachment, modeled as a regional, low-angle, ductile normal-fault zone of large displacement (refer to text). Note presence of Tertiary supracrustal rocks at same structural position across the entire exposed width of the detachment zone and the probably autochthonous nature of the Turtle Mountains. Severe geometric problems are encountered if mylonites from depths of 10 km or greater are juxtaposed by normal faulting with surficial upper-plate rocks, especially if the detachment "breakaway" zone lies just east of the Turtle Mountains (refer to text). B: Diagram showing large lateral displacements required to juxtapose surface Tertiary rocks with mylonites formed at considerable depth. This geometry is difficult to equate with actual relationships shown in A. C: Diagram showing much smaller lateral displacements if mylonitization develops at shallower depths. This configuration is more compatible with A.

rocks, once at depths greater than 10 km, would require two criteria to be met: (1) great *lateral* displacement (i.e., greater than 50 km; see Fig. 9B); and (2) continuity of the fault well to the northeast under the Colorado Plateau. Certainly, if the Tertiary volcanic rocks of the Mopah Mountains are para-autochthonous or autochthonous, then the allochthonous, northeast-transported, *coeval* volcanics above the Whipple-Buckskin mylonites could not have moved very far. The problem here is one of gross imbalance between supposed large lower- and upper-plate displacements along most of the structural section (Fig. 9), and the obvious minor to negligible displacements at the breakaway zone, a few miles to the west. The only way to preserve the shear zone model for this region is to significantly reduce the scale and displacement of the shear. One way to do this is to seriously reduce the depth of mylonite formation (Fig. 9C). Even then, however, the outcrop width of upper-plate Tertiary rocks juxtaposed against the mylonites would be quite restricted, unless the dip of the detachment fault remained extremely flat (just beneath the volcanic cover).

Applying the rotated and flattened detachment fault model of Davis, (1983) to the cross section of Figure 9 by restoring the 100-km-long detachment surface to an initial 45° dip (see Fig. 3 of Davis, 1983) would require that the original normal fault be continuous through the entire crust. Certainly no evidence for rocks this deep can be gleaned from lower-plate exposures across this entire line of section. The only alternative for the model is for the fault to lengthen substantially as it becomes flattened by extension.

6. Unique Upper-Plate Lithologies

In immediate proximity to the exposed mylonitic rocks of MCC, there occurs a variety of unique upper-plate Tertiary rocks not found to this extent elsewhere. This spatial correspondence would suggest an in situ relationship between upper and lower plates during extensional deformation and preclude their displacement laterally from any significant distances. The upper-plate rocks in question consist of red, chaotic, coarse breccias, conglomerates, and mafic volcanic-flow rocks. The volcanics (and sediments?) have suffered extreme potash metasomatism and hematization, but protoliths are thought to have been alkali basalts (Kerrich and others, 1986). The age range for these tectonic supracrustal sequences is estimated to be about 25 to 16 Ma (J. Otton, 1986, personal commun.). Coeval, intermediate to silicic tuffs, flows, and finer-grained sedimentary facies were deposited in a widespread region surrounding the Harcuvar and Whipple MCC of western Arizona and southeastern California.

7. Miscellaneous Points

The regional shear zone model does not seem to explain adequately the following features of MCC: (a) their more than fortuitous spatial coincidence with approximately coeval Tertiary plutons; (b) their anomalous gravity highs (Cady, 1980); (c) the occurrence of basic dioritic dikes; (d) the regional linearity of the MCC within the Cordilleran framework; and (f) their regional setting amid large-scale antiform–synforms and transverse, east-west–trending discontinuities.

One metamorphic core complex specifically illustrating some of the criteria discussed above is the Picacho MCC located midway between Tucson and Phoenix, Arizona. Core complex relationships in the Picacho MCC have been briefly alluded to by Rehrig and Reynolds (1980), Davis (1980), Banks (1980), and Rehrig (1982). More recently, unpublished field data accumulated by S. B. Keith and this writer argue against any middle Tertiary, *large-scale* crustal shear zone of major displacement. Pertinent relationships are shown in Figure 10 and key observations are summarized as follows:

(1). Part of the mylonitization is middle Tertiary in age, as evidenced by deformed early Miocene granodiorite pluton and felsite intrusive bodies. The pluton yields a Rb-Sr whole-rock age of 24.4 Ma, and felsites are dated at 25.9 Ma by whole-rock K-Ar (Rehrig, 1982). Aphanitic and in places porphyritic textures of the mylonitic felsite sills and dikes suggest ductile strain at fairly shallow depths. The main pluton is similar to other epizonal, holocrystalline bodies in the region. The zone of mylonitization and detachment faulting occurs at or just above the intrusive roof to the pluton. Therefore, if the pluton was shallowly emplaced, as is evidenced, then mylonite fabrics were also generated at this relatively high level.

(2). Exceptionally well-exposed contacts of the felsite dikes with Precambrian (1.4 Ga) quartz monzonite (Oracle Granite) within the mylonite zone reveal unsheared chilled margins to the high-level dikes, yet both felsite and Precambrian host rock are intensely mylonitic (Fig. 10C). In detail, the Oracle Granite appears to be the most deformed.

(3). In the lower plate, there are exposures that display gradational and conformable change upward from ductilely deformed mylonite to brittle breccia. Some of the upper part of this gradation is characteristic of what I call "transitional" mylonite, that texturally is transitional between ductile and brittle modes of intercrystalline deformation. The conformable, untruncated nature of the foliation through this interval supports the concept of a strain continuum from ductile to brittle conditions through a rather narrow (100 m) vertical range.

(4) The presence of a well-exposed middle plate (Fig. 10C) of brecciated and thoroughly propylitized granite derived from a generally nonmylonitic protolith establishes that chloritic breccia formed above lower-plate mylonite. In most MCCs, the detachment-related chloritic breccias are described as forming directly from the destruction of preexisting mylonitic rocks of the lower plate. The brecciated middle plate contains flat layers and lenses of microbreccia as well as complex deformational textures indicative of multiphased brittle to ductile strain transitions.

(5). The felsite dikes and sills are found in both lower and middle plates. In the lower plate, these bodies are mylonitic; in the middle plate, they are merely brecciated. This intrusive "pinning" suggests that ductile and brittle deformations were somewhat synchronous and were not widely separated in space. As previously indicated, both styles of deformation probably occurred at relatively shallow depths.

(6). There is a distinctly earlier phase of mylonite development at Picacho which is closely related to 2-mica and muscovite-garnet–bearing granite and pegmatite magmatism similar to that dated as Paleocene to Eocene (late Laramide) elsewhere in Arizona (Keith and Reynolds, 1981; Keith, 1982). These granites clearly cross cut extremely well-developed mylonitic rocks believed derived from Precambrian Oracle Granite. Some of the muscovite granites are overprinted by variable intensities of subtle foliated fabrics which are expressions of either Eocene or Miocene tectonism. The middle Tertiary (22–25 Ma) intrusion and detachment-related mylonitization were thus overprinted upon a major mylonite zone of flat fabric in the basement rocks of this area. This polyphase deformational history may apply as well to the neighboring Catalina–Rincon–Tortolita MCC, where rocks and fabrics are generally quite similar (Keith and others, 1980).

The observations at Picacho yield some inferences of importance to a MCC model; similarities exist with other MCC (i.e., Wilkens and Heidrick, 1982). Deformation within the mylonite detachment zone appears to have been a relatively in situ, Tertiary process with respect to the site of earliest Miocene intrusive activity. No great lateral or vertical displacements are indicated between mylonite and detachment breccias. The process involved development of a brittle, fragmented, chloritic breccia a relatively short vertical distance (i.e., <100 m) above the ductile mylonitic rocks. The complex localized mixing of breccias in mylonitic plate and *minor* mylonitic fabrics in chloritic breccia plate suggests a generally consanguineous relationship between mylonitization and detachment, one in which the strain oscillated between ductile and brittle conditions during a protracted extensional event. Ultimately, brecciation merely outlasted ductile conditions. As the syndeformational "riveting" of lower and middle plates by the felsites indicates juxtaposition during the extensional event, then surely mylonites, chloritic breccias, and detachment faults were all closely related under quite shallow in situ crustal conditions.

The Picacho example may be illustrative of what is seen elsewhere as detachment faults which are not underlain by mylonitic rocks (see Fig. 1D). Note from Figure 10 that the upper-plate volcanic rocks and subjacent detachment fault at Picacho rest upon the brecciated, but generally nonmylonitic, middle plate with mylonites some distance below. If such middle plates exist elsewhere in greater thickness, then the comparison between Picacho and Figure 1 would be useful in explaining some nonmylonitic detachment faults.

A
ARIZONA
P
T
P = PHOENIX
T = TUCSON
Ip
NEWMAN PEAK
Ti
A
A'
(FIG. 10B)
PHOENIX
upper plate Tv
DETACHMENT OUTCROP (FIG. 10 C)
INTERSTATE 10
Tv
PICACHO PEAK
45
20
50
UPPER PLATE 20-22 M.Y. VOLCANIC & CLASTIC ROCKS
DETACHMENT ZONE MULTIPLE FAULTS & BRECCIAS
LOWER PLATE MYLONITIC PRECAMBRIAN, LARAMIDE & MID-TERTIARY CRYSTALLINE ROCKS
Ti: 24 M.Y. GRANODIORITE STOCK
LOW-ANGLE FOLIATION & LINEATION
0 1 mi 1.7 km
TUCSON
B
West
A
East
A'
PICACHO DETACHMENT FAULT(S)
FIG. 10 C
1
2
3
4
5
6
6a
0 1 mile 1.7 km
Mylonitic fabric in 3
Mylonitic fabric in 6

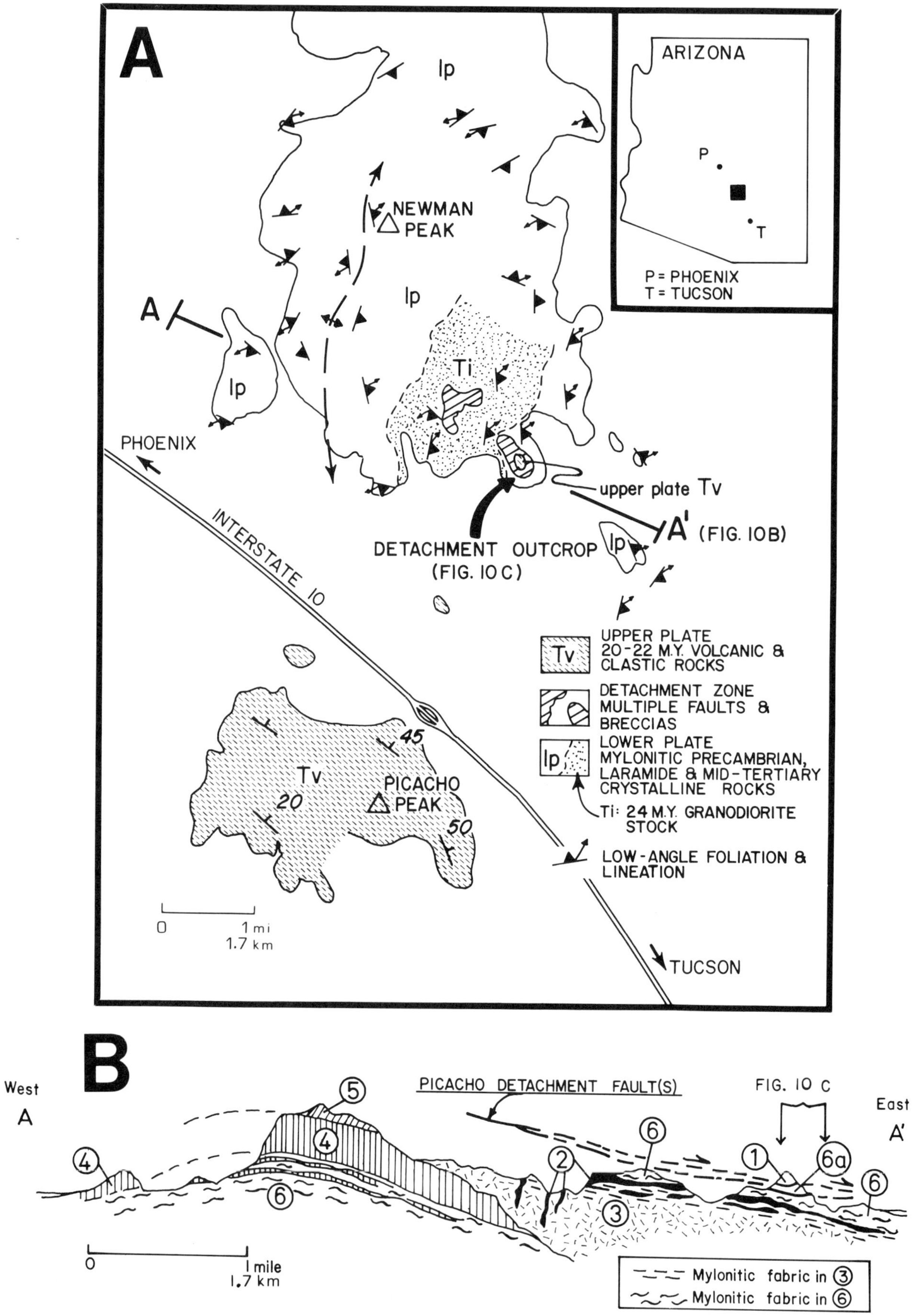

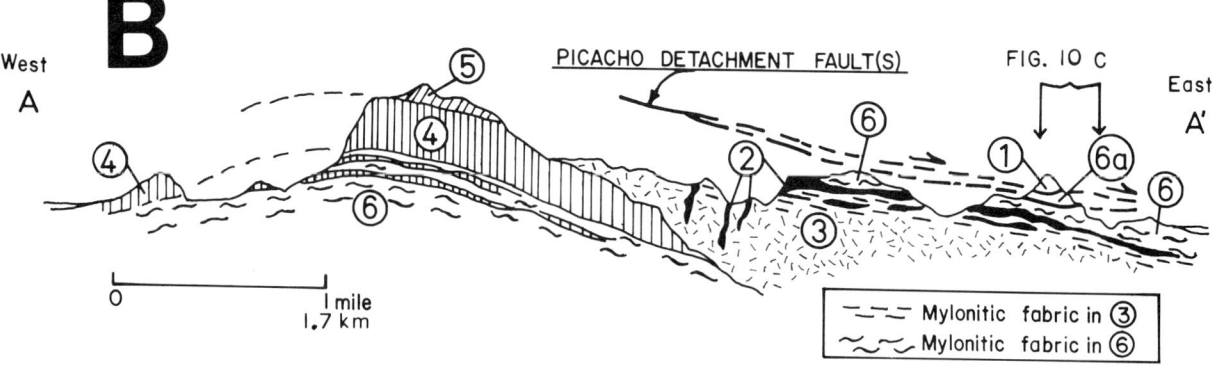

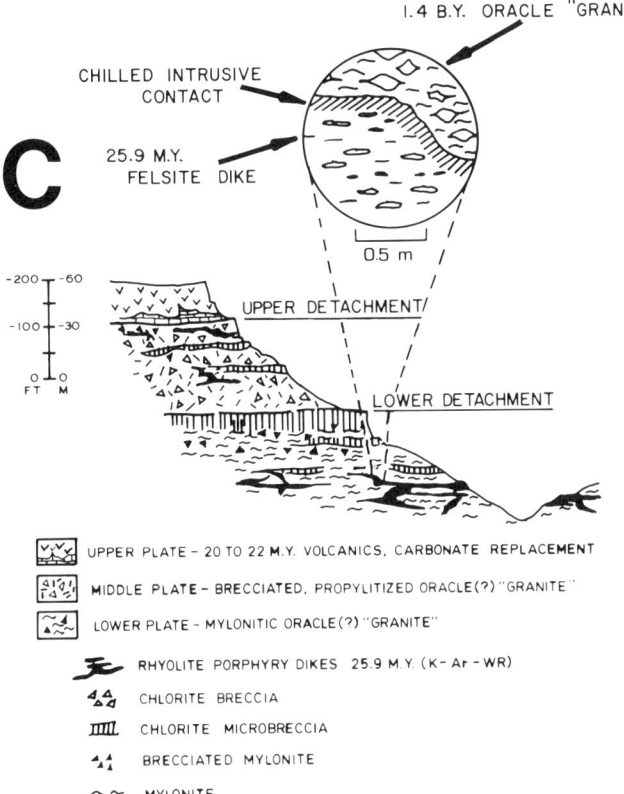

Figure 10. Relationships at the south end of the Picacho MCC in southern Arizona (Keith and Rehrig, unpubl. mapping; Shafiqullah and others, 1976; Davis, 1980). A (previous page): Tectonic map of the southern Picacho Mountains, showing distribution of lower- and upper-plate rocks and location of outcrops of the Picacho detachment zone. B (above): Schematic cross section across the complex showing relationship of middle Tertiary mylonite-detachment zone at top of early Miocene pluton. Pluton is intruded into a major, flat zone of probably late Laramide deformation and muscovite-garnet–bearing sill intrusion similar to that dated by U-Pb zircon as Eocene (50 to 40 Ma) in the neighboring Catalina Mountains to the south. Explanation to the cross section: (1) Miocene volcanics (22–20 Ma) (2) Alaskite and felsite dike/sills. (3) 24–25 Ma granodiorite pluton. (4) Eocene (?) 2-mica granite; slightly to moderately mylonitic throughout; becomes more muscovitic toward top of sill. (5) Eocene (?) pegmatite and muscovite granite sill complex intruding 4 and dark, mylonite schist above. Schist may be derived from 6. Sill complex is syn- to post-kinematic with respect to some of the mylonitization. (6) Mylonitic, 1.4 Ga Oracle Granite; at 6A, granite is part of chloritic, brecciated, middle plate of the Picacho detachment zone. C (right): Detailed cross section of the middle Tertiary detachment zone including upper, middle, and lower plates. Middle plate is predominantly nonmylonitic Oracle Granite (chloritic breccia) intruded by brecciated felsite dikes and sills which are strongly mylonitic in lower plate. Inset enlargement shows detail of intrusive contact of felsite dikes within mylonite zone. Note *unsheared* yet intensely mylonitic fabrics.

The preceding discussion should bring into focus the difficulty in assigning total predominance to either pure- or simple-shear mechanisms for MCC or their setting of regional extension. If anything, the examined data suggest that both mechanisms were operative. The likelihood of discrete low-angle shear surfaces—some with mylonites, some without—is acknowledged; however, I have made a case that some are small-scale, high-level structures. It is also argued that an indistinguishable mixture of shear and stretching fabrics can be expected in an extensional system controlled principally either by simple shear or by nonuniform, layer-parallel, pure shear (Fig. 6).

In the following section, some thoughts regarding mechanical processes responsible for MCC development are described. The model is largely based on new data from the chloritic breccia zone which occurs between ductile mylonites of the lower plate and dismembered upper-plate rocks.

THE CHLORITIC BRECCIA ZONE AND ITS RELEVANCE TO A METAMORPHIC CORE COMPLEX MODEL

I believe that without exception, the Cordilleran MCC of the western United States are marked by the characteristic chloritic breccia which occurs at the very top of the mylonitic lower plate. This zone of propylitized, brecciated rock was originally described as a kind of fault breccia derived from break up of the ductilely deformed tectonites by detachment faulting. To those advocating a shear zone model for the MCC, the chloritic breccia is still regarded as the brittle fault rock equivalent of the deeper mylonites, developed along the detachment surface at its shallower updip levels (Davis and others, 1983; Wernicke, 1985; Davis and others, 1986).

New data on the breccia and its relationships with upper- and lower-plate rocks, however, impart a unique significance to the breccia, one I regard as critical to a tectonic model. The chloritic breccia is generally best developed where it is superposed upon mylonitic rocks. Where the tectonized rocks are not present along a detachment fault, the propylitized interval is either absent or is poorly developed (Phillips, 1982; Lyle, 1982; McClelland, 1982; R. K. Dokka, 1983, personal commun.). Nowhere is this better demonstrated than in the Whipple MCC of southeast California. In this complex, the chloritic breccia zone is exceptionally well developed in contact with the mylonites but is absent beneath the detachment fault of Savahia Peak west of the mylonitic front (Phillips, 1982) (see Fig. 9). In the Riverside Mountains just south of the Whipple complex, a nonmylonitic detachment fault has essentially no chloritic breccia. There appears to be a close spatial coincidence between the mylonite and the breccia.

Mesoscopic and microscopic textures of the chloritic breccia are not indicative of a fault breccia. At all scales, there is an intricate, angular, interlocking fabric of fractures (Phillips, 1982) that show no rolling or rounding as might be expected by brecciation along a shear surface. Instead, concomitant mineralization and fluidized or entrainment textures in the breccia are clearly of hydrothermal origin. Near the top of the mylonite interval, incipient brecciation is first noted as microscopic cracks, only millimeters in length, which commonly show consistent orientation and are mineralized with chlorite (Rehrig and others, 1986a). The disconnected and small-scale nature of these hydrofractures strongly implies an in situ source for the fluids. In the chloritic breccia, the lengths and overall intensity of hydrofracturing progressively increase upward until a pervasive, interlocking stockwork of fractures is developed that is mineralized with chlorite, epidote, quartz, ± albite, ± calcite, ± sparce sulfides. Commonly, a vertical mineral zoning occurs with predominant epidote, quartz, and albite above chlorite (Rehrig and others, 1986a). Generally, the rocks become progressively lighter in color upward in the chloritic breccia column. At the top of the column, a well-indurated, dense, chloritic microbreccia exhibits the same kind of interlocking, angular fragmentation on a scale down to 0.0004 mm (Phillips, 1982, Fig. 5). The character of the overall breccia argues for an upward increase in fluids at the top of the lower-plate mylonites. These fluids were apparently derived from deuteric sources within the lower plate. The buildup of high fluid pressures culminated in the formation of microbreccia in a superplastic(?) fluidized medium.

Of equal importance is the relatively recent recognition of middle plates containing nonmylonitic rocks above a detachment fault (i.e., Wilkens and Heidrick, 1982). Fluids responsible for chloritic brecciation have crossed over the lower detachment boundary of these plates, totally affecting middle plate rocks (see previous description of Picacho MCC). The significance here is that fluids suspected of having been derived from lower-plate rocks (i.e., the ductile tectonites) have traveled across a detachment surface into nonmylonitic, upper-plate-like rocks. If the fluids responsible for the chloritic breccia originated from dewatering of the mylonites during their ductile deformation, as textural evidence and study of similar rocks (Kerrich and others, 1980) suggest, then this fluid communication across the detachment establishes lower-middle-plate proximity during detachment faulting.

Detachment breccias have two alteration phases: one is chloritic (propylitic), the other, a later phase, is characterized by intense iron oxide (hematite) and carbonate (Phillips, 1984). Both mineralization phases are represented in metallic mineral deposits associated with detachment faulting (Rehrig, 1984). The oxiderich alteration, accompanied by extremely enhanced potash, manganese, and calcium content, has pervasively affected upperplate Tertiary volcanic and clastic rocks and its diminishing effects are noted downward through the chloritic breccias (Kerrich and others, 1986). Upper-plate volcanics display this metasomatism most intensely where the rocks are in direct contact with the mylonitic detachment. With distance away from the detachment zone, the alteration diminishes (Rehrig, 1984; E. Brooks, 1984, personal commun.). These observations provide more evidence for detachment-related fluid regimes between upper and lower plates.

Preliminary geochemical comparisons among undeformed protolith (1.4 Ga granite), its mylonitized and chloritic brecciated equivalents, and altered/unaltered upper-plate volcanics (Rehrig, 1984) further argues for fluid transmissibility throughout the ductile to brittle deformed detachment zone. Elemental losses and gains from protolith to mylonite appear qualitatively matched by compensating gains and losses within the chloritic breccias. Oxygen isotopic data from the overall deformed zone suggest that a single-source fluid of metamorphic or magmatic origin derived from the lower plate was responsible for the upward increasing alteration and brecciation (Kerrich and others, 1984; Kerrich and others, in prep.). This fluid cooled rapidly upward, exchanging significantly with the altered rocks yielding progressively heavier isotopic oxygen.

My thoughts regarding the ductile-brittle deformation zones of MCC (Fig. 11) are that the zones rely substantially upon this fluid regime, which indicates shallowness and relatively in situ geometric constraints for the various tectonic plates during deformation. To restate: if chemical changes and fluid evolution are a direct result of lower-plate deformation, and if these changes are manifested in brecciated detachment plates above, then: (1) all plates must have been roughly juxtaposed during deformation; (2) there are no great thicknesses of "missing" rocks between lower and upper plates; and (3) mylonitization can develop at shallow crustal levels.

One problematic aspect of the layer stretching model for MCC (Rehrig and Reynolds, 1980) is that it does not adequately explain the observed gradational decrease in intensity of mylonitization with depth. A flattened and stretched layer sandwiched between nonductile zones should create a sheared contact at its base as well as at its top. This basal shear is not seen in the complexes. The process described below and illustrated in Figure 11 may explain this problem by permitting a lower gradational strain boundary to the mylonite zone and upward intensification of mylonitization toward the breccia and detachment surface.

The key ingredient to the MCC model proposed herein is the development of a fluid phase that facilitated ductile strain (mylonitization) and subsequently catalyzes formation of the chloritic breccia. The role of fluids in core complex genesis was first mentioned by Davis and Coney (1979), and stressed by Rogers (1981), but has since been forgotten. As discussed above, relatively large cumulative volumes of fluids may have originated from the process of mylonitization and detachment. Because of the correlation of nearly synchronous Tertiary plutons with some MCC terranes, magmatic fluids may have provided additional sources of water in some areas.

Experimental work has now established that structural water is an important catalytic agent in weakening quartz and other silicate minerals to ductile strain (Nicholas and Poirier, 1976; Kerrich and Allison, 1978; Bell and Etheridge, 1976). Weakening occurs by lowering the temperature at which grain boundary diffusion can cause pressure solution, by effectively reducing confining pressures and by reducing yield strengths of minerals due to hydraulic weakening (Griggs, 1967; Sibson, 1977). For quartz, the main ductile mineral of mylonites, plasticity can be attained under relatively low pressure and low-temperature greenschist conditions (Sibson, 1977). Recent studies (i.e., Kerrich and others, 1980; Phillips, 1982) continue to supply data indicating that plastic or ductile conditions can be attained in

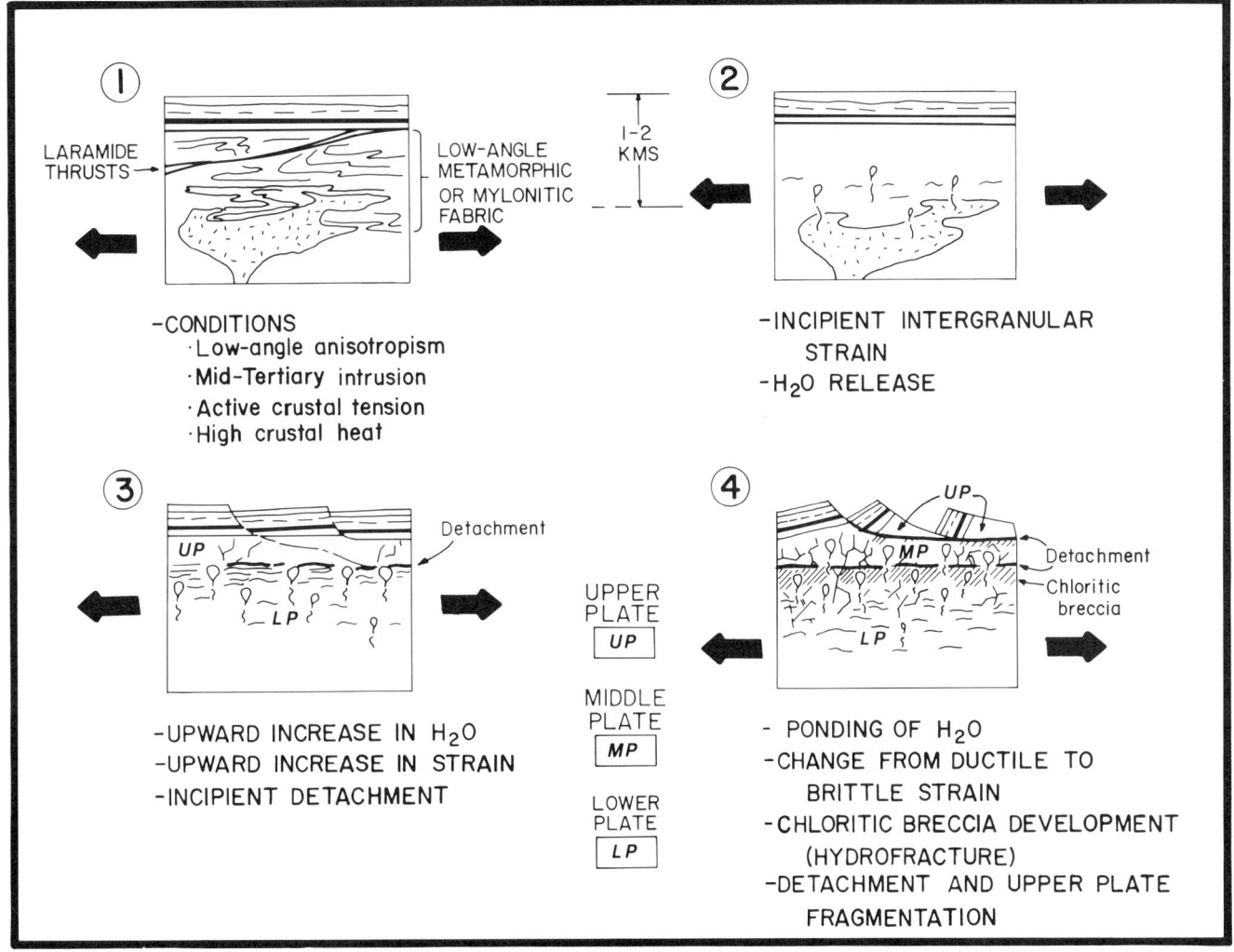

Figure 11. Postulated evolution and development of mylonitic detachment zones of metamorphic core complexes in domains of low-angle anisotropism, high crustal heat, Tertiary intrusion, dynamic extensional stress fields, and increasing water contents. Incipient release of fluids by intergranular strain softens the rock progressively upward to enhance ductile fabric development leading to detachment between ductile and brittle plates and formation of chloritic breccia (refer to text). Note that the process as envisaged can occur in situ at quite shallow depths (in upper parts or roofs of epizonal Tertiary plutons).

upper crustal, relatively low temperature and pressure (i.e., 250°C and <5 km depth) domains, if strain rates are sufficiently low and pore fluid pressures are sufficiently high.

Further microstructural research by Kerrich and others (1980, 1981) has proven that stress-induced microfracturing and intergranular deformation in mylonitic rocks can liberate small quantities of structural fluids which probably enhance lattice diffusion within crystals and catalyze mineral reactions. These fluids and the resulting mineral alterations concentrate along stress corrosion cracks and grain boundaries. The accumulation and upward migration of these fluids alters preexisting minerals, forms new ones, and facilitates rock deformation, and is the overall process envisaged within the Tertiary detachment zones of the MCC.

The upper and most deformed mylonites are thus generated in a relatively water-rich environment, evidenced by water contents of mylonitic rocks which are higher than those of undeformed protoliths (Sibson, 1977). Within the upper mylonite section, substantial mineralogical destruction and reconstitution are noted, the most common being liberation of secondary silica, breakdown of ferromagnesium minerals, and alkali readjustments in feldspars (Rehrig and others, 1986a). These chemical changes, which correspond to more intense deformation, further substantiate fluid mobility within the ductile to brittle detachment zone. Geochemical modifications have also been noted by other workers (Keith and Reynolds, 1980; Bridgewater, 1979; Banks, 1980; Kerrich and others, 1977, 1981).

Cordilleran MCC thus originated in areas controlled by the

following factors: (1) high heat flow or intrusion, (2) regional extension, (3) previously formed, low-angle anisotropic fabrics, and (4) presence of metamorphic or magmatically supplied fluids. Ductile deformational fabrics may have initially developed along some relatively continuous strain discontinuity or zone of flat or low-angle geometry. Penetrative strain induced intercrystalline exsolving of fluids on a microscopic scale. Rock softening began and compounded both deformation and water accumulation in an upwardly intensifying mylonitic zone. The zone evolved upward with time and resulted in relative displacement between upper and lower plates across the ductile-brittle strain boundary. The thick zone of deformational fabrics need not have resulted, however, from excessive overall displacement across the detachment zone. Upper-plate rocks were extended by brittle block rotation and tilting above the zone of ductile mobility below (Fig. 11).

The process of chloritic brecciation, which is superimposed upon mylonitization, is visualized in either of two ways. (1) Continued water buildup at relatively shallow levels from ductile deformation in the lower plate causes a switch from ductile to brittle conditions accompanied by extensive retrograde metamorphic alteration. In essense, brecciation becomes an explosive, hydrofracturing phenomenon. Experimental and geologic evidence for this type of deformational transition was cited by Rutter (1976) and Ashby and Verrall (1977). Their data indicated that changes from ductile to brittle conditions can occur through a wide range of depths and temperatures merely by increasing internal fluid pressures in the rocks. (2) Upon uplift and cooling, the hydrated mylonites release structural water which accumulates or ponds at the detachment horizon and leads to hydraulic fracturing, brecciation, and propylitic mineral reactions. Experimental studies suggest that the solubility of H_2O in nominally anhydrous minerals such as quartz increases during prograde metamorphism. During post metamorphic uplift and temperature decrease, water release from these minerals is postulated (Spear and Selverstone, 1983). If minute quantities of water derived in this manner are allowed to migrate through the rock and pond in any receptive horizon of relatively small overall volume, significant amounts of water can be accumulated. The water is then available for extensive retrograde alteration reactions and for brecciation (Fig. 11).

The hypothesis that in situ waters are the catalytic agents for generation of the chloritic breccia is helpful in explaining the widespread, essentially indigenous character of the propylitic alteration associated with mylonitic detachment zones. Because of this two-dimensional continuity, it is difficult to imagine a fluid source from any site-specific foreign location. The concentrated water at the detachment horizon may explain the geochemical variations noted previously and is compatible with the oxygen isotopic data. Finally, the water "flushing" in the chloritic breccia zone may explain the incredible temperature gradients existing between lower and upper plates. As posed by Wernicke (1985), a major problem with any generally in situ model is how to keep the upper plate from heating up in contact with the hot (400–500 °C), ductile lower plate. The relatively high water contents within the chloritic breccia intervals might help to dissipate or carry away this heat.

The hypothesis of a causal continuum between mylonites and detachment breccias in the presence of water is proposed with full regard to the generally acknowledged facts that (1) breccia follows mylonite in sequence, (2) a bona fide shear surface (detachment fault) overlies the chloritic breccia, and (3) there is abundant textural evidence for uniaxial simple shear. Hydrofracturing and brecciation generally occur *after* mylonites develop and release their structural waters. This time interval, however, may be very short. A major discontinuity or profound shear must develop between the intensely deformed mylonite zone and surficial upper plate. Within the zone of mylonites, layer-parallel ductility variations are extreme (Fig. 6), even over intervals measured in centimetres. Thus, intra-layer shear is inevitable, even if the strain is dominantly one of ductile extension (pure shear). It would therefore be expected that in the general zone of detachment between regimes of extremely contrasting differential extension ample evidence for simple shear could be documented. If stretching was directed predominantly in one direction from some fixed position, then aspects of unidirectional shear would dominate.

Summary Statement

Before closing this section, it should be emphasized that there is no intent here to eliminate the concept of ductile, low-angle, normal-fault zones that are expected to develop in an extending orogen. I have, however, tried to make the point that these shear zones are in many places much smaller in scale than those suggested in the literature. In addition to the standard configuration for such shears (Wernicke, 1981, 1985; Davis and others, 1983), there are other mechanisms for generation of flat detachment structures at relatively shallow crustal levels (see Fig. 1). If the crust is thermally and hydraulically softened in a dynamic tensional stress field, then ductile stretching should be at least locally anticipated at these levels; that is the keynote of much of the preceding discussion. My emphasis on shallowness should not negate the actuality of detachments and mylonites developed at substantially deeper levels. Evidence for these features at depths greater than 6 km has been reported by Proffett (1977), Gans and Miller (1983), and Howard and others (1982a). Note that if rocks actually deform by a predominance of pure shear, then several of the mechanisms depicted in Figure 1 could easily lead to MCC-type features. The preceding genetive discussion for MCC deals with the specifics of fabric generation in a zone of extreme tectonism at the contact between ductile and brittle crustal layers, but it could apply to other mechanisms (Fig. 1). It could even be accommodated by the simple shear zone model, with important qualifications: (1) the shear would have to be quite shallow and flat, and (2) overall displacements between lower and upper plates would not be large. (3) The shears need not be of great regional continuity transmitted through the entire crust.

PLATE TECTONIC SETTING

The metamorphic core complexes are but one element within the extended western Cordillera. The complexes, distributed in a rather narrow zone, fit within a huge region of overall antiformal geometry marked by variably tilted mountain blocks, subdomains of antiformal-synformal configuration, and unusual transform discontinuities (Stewart, 1979). One obvious question, for which there is no easy answer, is whether there are MCC-like features in the subsurface throughout the orogen. From the character of recent, high-resolution seismic data (Hamilton, 1982; Anderson and others, 1983; Allmendinger and others, 1983; Smith, 1983; Frost and Okaya, 1986), flat mylonitic zones and detachment discontinuities appear likely in the subsurface. These structures would certainly constitute a means of extending the surficial brittle rocks throughout the province. The uplifted, intruded, and structurally attenuated MCC may be localized analogs to the extended western Cordillera.

Existing literature is vague regarding ultimate tectonic causes for the Tertiary extensional activity in western North America. Concerning the earlier Laramide orogeny, there has been much said of its plate tectonism in a setting of extremely high convergence rates between the North American and Farallon plates and flattened subduction (Coney and Reynolds, 1977; Lipman, 1981; Dickinson, 1981; Keith, 1982). Plate tectonic explanations for the Eocene to Miocene extensional tectonism, however, have been limited to those of decreased subduction rates and progressive steepening of the Benioff zone (see preceding references). One problem with the existing plate tectonic scenario can be stated as follows: A comparison of subduction histories during *early* Laramide and middle Tertiary time in the southwestern Cordillera yields similar Benioff zone geometries and magmatic sequences. Tectonism, however, was radically different; the Laramide exhibited compression and moderate crustal shortening; the middle Tertiary resulted in extreme extension and crustal dilation. A similar disparity exists between middle to Late Cretaceous and Eocene tectonism in the northwestern United States.

Any explanation for the Cordillera extensional orogeny must address several of its most important characteristics which include (1) a dynamic, forceful pull-apart and distension of the crust; (2) abnormally high crustal heat flow (geothermal gradients); and (3) widespread, relatively shallow emplacement of plutons and surface buildup of enormous quantities of volcanic material. Through all but the eastern margin of the province, the magmatism is generally calc-alkaline and of overall intermediate or andesitic composition. One unusual characteristic of its generation is a nearly simultaneous burst across the entire 1000-km width of the Cordillera (at least in the southwest and northwest United States) during a relatively brief 10 m.y. interval.

The following plate tectonic model, although speculative, is applied to the southwestern Cordillera (southern California, Arizona, New Mexico). It may have application to the northwest United States, where a remarkably similar history occurred some 20 m.y. earlier. During the Paleocene-Eocene (~50–40 Ma), calc-alkalic, metaluminous magmatism ceased throughout much of the southwest (the Laramide magmatic gap; Damon and Mauger, 1966). A newly recognized pulse of unique, peraluminous, 2-mica-garnet–bearing granitoids is seen to fill the gap (Keith and Reynolds, 1980; Keith, 1982). This s-type plutonism is thought to be associated with exceptionally low-angle subduction where the subducted plate was too shallow and the mantle wedge too thin (or nonexistent) to facilitate normal, Andean-type magmatism. The period of what we can call middle Tertiary magmatism started no earlier than 40 Ma (Elston, 1984), but the well known Coney and Reynolds (1977) magmatic plot shows few magmatic dates before 30–35 Ma. The very flat trace of the lower part of the Coney-Reynolds curve reflects the nearly synchronous nature of magmatic activity across the 1000-km width of the southern Cordillera. The duration of this magmatic burst was about 10 to 15 m.y. It correlated closely with crustal extension and was calc-alkaline and metaluminous, nearly indistinguishable from magmatism of the classic Laramide orogeny (75–50 Ma) (Rehrig, 1981, 1982). In brief, something relatively catastrophic had to initiate such widespread crustal thermal and extensional orogenic activity. A key constraint of the process was magma genesis no different from that of the earlier Laramide, compressive, Andean-like orogeny.

To many, the middle Tertiary orogeny was triggered by rapid decreases in plate convergence rates leading to actual cessation of subduction along the southwest margin of North America, perhaps as early as 25–30 Ma (Atwater, 1970; Coney and Reynolds, 1977; Engebretson and others, 1982; Elston, 1984). To this author, the near juxtaposition of the cold, subducted Farallon slab to the base of the crust for 10 m.y. preceding the middle Tertiary episode was important in extinguishing metaluminous magmatism. After the sudden decrease or termination of plate convergence after about 40 Ma, the subducted slab, lacking the dynamic lateral forces to keep it shallow, may have segmented at its western edge (Fig. 12). The cold slab would then sink through the hotter upper mantle, leading to widespread dehydration at some level (>100 km). This water release, coupled with ingress of hot asthenosphere between sinking slab and crust, might have led to massive lower crustal(?) melting observed empirically across the entire Cordillera along many latitudes.

The lateral and upward infusion of such quantities of mantle material and consanguinous melting of lower crust should have led to crustal instability (thinning and softening) and even mantle diapirism (Fig. 12). If the asthenospheric migration occurred in one lateral direction, it would have imparted strong subhorizontal stresses to the crust. Divergent motions would have led to crustal pull-apart and the array of extensional structures described in this paper. There seem to be distinct domains within the Cordillera where unidirectional pull apart was focused. As extensional fragmentation and overall crustal thinning evolved, access for progressively shallower diapirism and convection might have developed.

It is not difficult to envisage in this setting local environ-

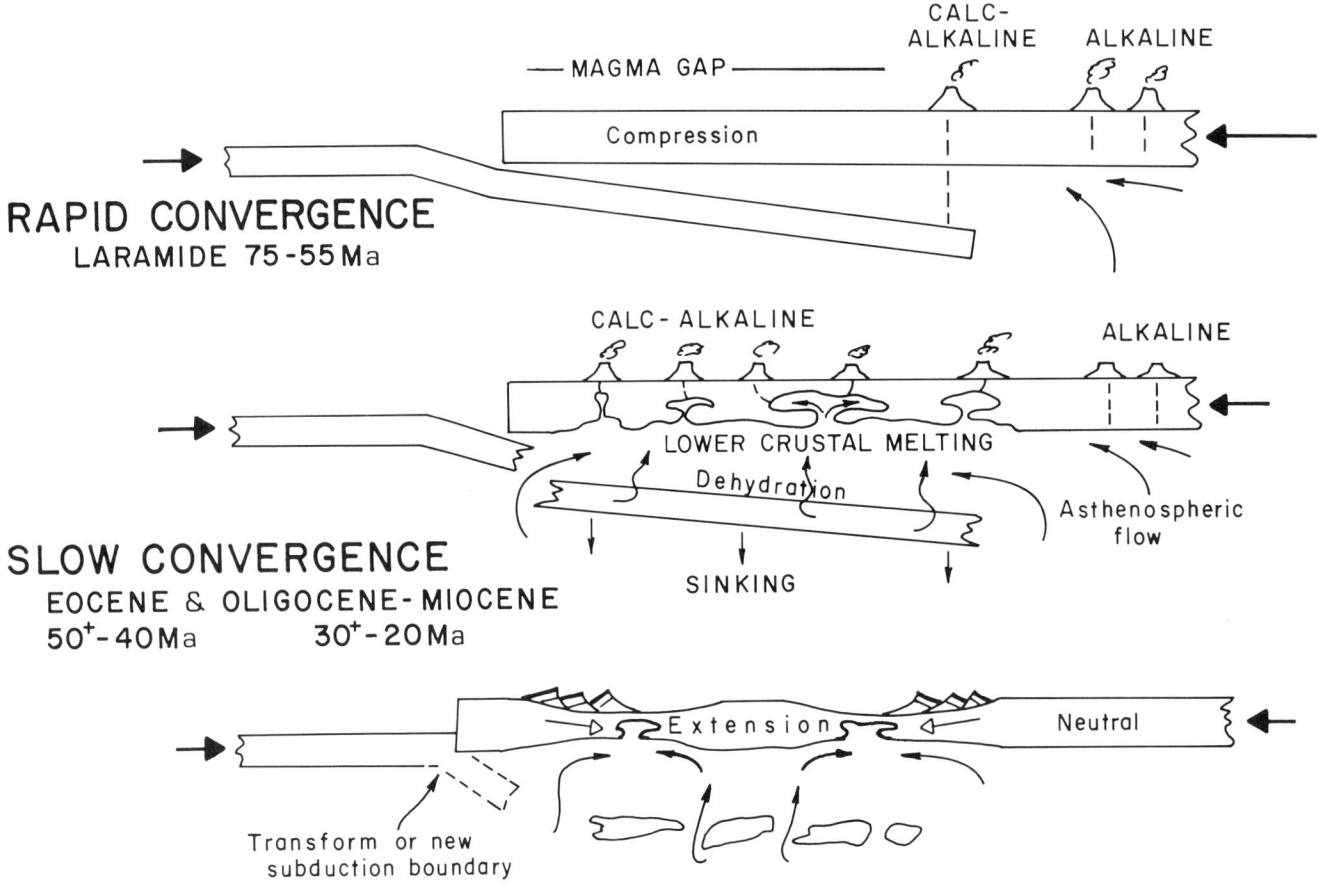

Figure 12. Proposed plate tectonic mechanisms for generation of the Eocene (northwest) and Oligocene–Miocene (southwest) extensional events. Tectonic process is keyed to rupture and sinking of the extremely shallow Laramide subduction zone upon rapid decrease in plate convergence rates 40 to 35 Ma. Subduction probably ceased altogether in the southwest about 30 Ma, resulting in asthenospheric inflow and diapirism into crust, which led to widespread magmatism, crustal thinning and softening, and continental extension.

ments of major dilational pluton emplacement, thermal softening, high-level crustal stretching, and even ductile, shallow-dipping, normal-shear zones. Ample opportunity would have existed for the evolution of subhorizontal fabrics and structures in the crust which facilitated Cordilleran extension. Perhaps the abundant, late-stage mafic (diorite) dikes which intrude the MCC are a manifestation of the basic, lower-crustal upwellings which propelled crustal distension. Additionally, the enigmatic gravity highs that characterize a number of MCC may be caused by substantial volumes of this basic magmatic material, which promoted ductility and filled the thinned zones.

ACKNOWLEDGMENTS

Credit should be given to CONOCO, Inc. and Phillips Petroleum Company; it was in their employ (1975–1982) that I gained much of my experience with metamorphic core complexes. Further thanks are extended to the editor of this publication for his patience, and to critical reviewers E. DeWitt and J. H. Stewart of the U.S. Geological Survey. Ed DeWitt especially helped to greatly improve early versions of the paper.

REFERENCES CITED

Allmendinger, R. W., Sharp, J. W., Von Tish, D., Serpa, L., Brown, L., Kaufman, S., and Oliver, J., 1983, Cenozoic and Mesozoic structure of the eastern Basin and Range province, Utah, from COCORP seismic-reflection data: Geology, v. 11, p. 512–516.

Anderson, R. E., 1971, Thin skin distention in Tertiary rocks of southeastern Nevada: Geological Society of America Bulletin, v. 82, p. 43–58.

Anderson, R. E., Zoback, M. L., and Thompson, G. A., 1983, Implications of selected subsurface data on the structural form and evolution of some basins in the northern Basin and Range province, Nevada and Utah: Geological Society of America Bulletin, v. 94, p. 1055–1072.

Anderson, T. H., Silver, L. T., and Salas, G. A., 1980, Distribution and U-Pb isotope ages of some lineated plutons, northwestern Mexico, in Crittenden, M. D., Jr., Coney, P. J., and Davis, G. H., eds., Cordilleran metamorphic core complexes: Geological Society of America Memoir 153, p. 269–286.

Armstrong, R. L., 1972, Low-angle (denudation) faults, hinterland of the Sevier orogenic belt, eastern Nevada and western Utah: Geological Society of America Bulletin, v. 83, p. 1729–1754.

Ashby, M. F., and Verrall, R. A., 1977, Micromechanisms of flow and fracture, and their relevance to the rheology of the upper mantle: Cambridge University Engineering Department Report Mat/TR30, 45 p.

Atwater, T., 1970, Implications of plate tectonics for the Cenozoic tectonic evolution of western North America: Geological Society of America Bulletin, v. 81, p. 3513–3536.

Banks, N. G., 1980, Geology of a zone of metamorphic core complexes in southeastern Arizona, in Crittenden, M. D., Jr., Coney, P. J., and Davis, G. H., eds., Cordilleran metamorphic core complexes: Geological Society of America Memoir 153, p. 177–216.

Bell, T. H., and Etheridge, M. A., 1976, The deformation and recrystallization of quartz in a mylonite zone, central Australia: Tectonophysics, v. 32, p. 235–267.

Berg, Lindee, Leveille, G., and Geis, P., 1982, Mid-Tertiary detachment faulting, and manganese mineralization in the Midway Mountains, Imperial County, California, in Frost, E. G., and Martin, D. L., eds., Mesozoic-Cenozoic tectonic evolution of the Colorado River region, California, Arizona, and Nevada: San Diego, Cordilleran Publishers, p. 298–312.

Bridgewater, D., 1979, Chemical and isotopic redistribution in zones of ductile deformation in a deeply eroded mobile belt: U.S. Geological Survey Open-File Report 79-1239, p. 505–526.

Brown, R. L., and Read, P. B., 1983, Shuswap terraine of British Columbia: A Mesozoic "core complex": Geology, v. 11, p. 164–168.

Cady, J. W., 1980, Gravity highs and crustal structure, Omineca crystalline belt, northeastern Washington and southeastern British Columbia: Geology, v. 8, p. 328–332.

Chapin, C. E., Chamberlin, R. M., Osburn, G. R., White, D. W., and Sanford, A. R., 1978, Exploration framework of the Socorro geothermal area, New Mexico: New Mexico Geological Society Special Publication 7, p. 115–129.

Chase, R. B., Bickford, M. E., and Arruda, E. C., 1983, Tectonic implications of Tertiary intrusion and shearing within the Bitterroot dome, northeastern Idaho batholith: Journal of Geology, v. 91, p. 462–470.

Colletta, B., and Angelier, J., 1982, Sur les systemes de blocs failles bascules associes aux fortes extensions: etude preliminaire d'exemples ouest-americains (Nevada, USA et Basse-California, Mexique): Paris, Académie des Sciences Compte Rendus, v. 294, p. 467–469.

Compton, R. R., 1980, Fabrics and strains in quartzites of a metamorphic core complex, Raft River Mountains, Utah, in Crittenden, M. D., Jr., Coney, P. J., and Davis, G. H., eds., Cordilleran metamorphic core complexes: Geological Society of America Memoir 153, p. 385–398.

Compton, R. R., Todd, V. R., Zartman, R. E., and Naeser, C. W., 1977, Oligocene and Miocene metamorphism, folding and low-angle faulting in northwestern Utah: Geological Society of America Bulletin, v. 88, p. 1237–1250.

Coney, P. J., and Harms, T. A., 1984, Cordilleran metamorphic core coplexes: Cenozoic extensional relics of Mesozoic compression: Geology, v. 12, p. 550–554.

Coney, P. J., and Reynolds, S. J., 1977, Cordilleran Benioff zones: Nature, v. 270, p. 403–406.

Crittendon, M. D., Jr., Coney, P. J., and Davis, G. H., editors, 1980, Cordilleran metamorphic core complexes: Geological Society of America Memoir 153, 490 p.

Damon, P. E., and Bikerman, Michael, 1964, Potassium-argon dating of post-Laramide volcanic rocks within the Basin and Range province of southeast Arizona and adjacent areas: Arizona Geological Society Digest, v. 7, p. 63–78.

Damon, P. E., and Mauger, R. L., 1966, Epeirogeny-orogeny viewed from the Basin and Range province: Society of Mining Engineers, Transactions, v. 235, no. 1, p. 99–112.

Davis, G. A., Anderson, J. L., Frost, E. G., and Shackleford, T. J., 1980, Mylonitization and detachment faulting in the Whipple-Buckskin-Rawhide Mountains terrane, southeastern California and western Arizona, in Crittenden, M. D., Jr., Coney, P. J., and Davis, G. H., eds., Cordilleran metamorphic core complexes: Geological Society of America Memoir 153, p. 79–129.

Davis, G. A., Anderson, J. L., Martin, D. L., Krummenacher, D., Frost, E. G., and Armstrong, R. L., 1982, Geologic and geochronologic relations in the lower plate of the Whipple detachment fault, Whipple Mountains, southeastern California: A progress report, in Frost, E. G., and Martin, D. L., eds., Mesozoic-Cenozoic tectonic evolution of the Colorado River region, California, Arizona, and Nevada: San Diego, Cordilleran Publishers, p. 408–432.

Davis, G. A., Lister, G. S., and Reynolds, S. J., 1983, Interpretation of Cordilleran core complexes as evolving crustal shear zones in an extending orogen: Geological Society of America Abstracts with Programs, v. 15, p. 311.

Davis, G. A., Lister, G. S., and Reynolds, S. J., 1986, Structural evolution of the Whipple and South mountains shear zones, southwestern United States: Geology, v. 14, p. 7–10.

Davis, G. H., 1975, Gravity-induced folding off a gneiss dome complex, Rincon Mountains, Arizona: Geological Society of America Bulletin, v. 86, p. 979–990.

—— , 1980, Structural characteristics of metamorphic core complexes, in Crittenden, M. D., Jr., Coney, P. J., and Davis, G. H., eds., Cordilleran metamorphic core complexes: Geological Society of America Memoir 153, p. 35–78.

—— , 1983, Shear-zone model for the origin of metamorphic core complexes: Geology, v. 11, p. 342–347.

Davis, G. H., and Coney, P. J., 1979, Geoiogic development of Cordilleran metamorphic core complexes: Geology, v. 7, p. 120–124.

Davis, G. H., Anderson, P., Budden, R. T., Keith, S. B., and Kiven, C. W., 1975, Origin of lineation in the Catalina-Rincon, Tortolita gneiss complexes, Arizona: Geological Society of America Abstracts with Programs, v. 7, p. 602.

DeWitt, E., 1980, Comment on "Geologic development of Cordilleran metamorphic core complexes": Geology, v. 8, p. 6–9.

Dickinson, W. R., 1981, Plate tectonic evolution of the southern Cordillera, in Dickinson, W. R., and Payne, W. D., eds., Relations of tectonics to ore deposits in the southern Cordillera: Arizona Geological Society Digest, v. XIV, p. 113–136.

Dokka, R. K., 1981, Early Miocene detachment faulting in the central Mojave Desert, California: U.S. Geological Survey Open-File Report 81-503, p. 29–34.

Elston, W. E., 1984, Subduction of young oceanic lithosphere and extensional orogeny in southwestern North America during Mid-Tertiary time: Tectonics, v. 3, p. 229–250.

Engebretson, D. C., Cox, A. V., and Thompson, G. A., 1982, Convergence and tectonics: Laramide to Basin-Range [abs.]: EOS American Geophysical Union (Transactions), v. 63, p. 911.

Fox, K. F., Jr., Rinehart, C. D., Engels, J. C., and Stern, T. W., 1976, Age of emplacement of the Okanogan gneiss dome, north-central Washington: Geological Society of America Bulletin, v. 87, p. 1217–1224.

Frost, E. G., and Okaya, D. A., 1986, Application of seismic reflection profiles to tectonic analysis in mineral exploration, in Beatty, B., and Wilkinson, P.A.K., eds., Frontiers in geology and ore deposits of Arizona and the southwest: Arizona Geological Society Digest, v. XVI, p. 137–152.

Gans, P. B., and Miller, E. L., 1983, Style of mid-Tertiary extension in east-central Nevada, in Guidebook, Part I, Geological Society of America Rocky Mountain and Cordilleran sections meeting: Utah Geological and Mining Survey Special Studies, v. 59, p. 107–160.

Gans, P. B., Miller, E. L., McCarthy, J., and Ouldcott, M. L., 1985, Tertiary extensional faulting and evolving ductile-brittle transition zones in the northern Snake Range and vicinity: New insights from seismic data: Geology,

v. 13, p. 189–193.

Garmezy, L. G., 1984, Geology and geochronology of the southeastern Bitterroot Mountains [Ph.D. thesis]: State College, Pennsylvania State University.

Garner, W. E., Frost, E. G., Tanges, S. E., Germinario, M. P., 1982, Mid-Tertiary detachment faulting and mineralization in the Trigo Mountains, Yuma County, Arizona, *in* Frost, E. G., and Martin, D. L., eds., Mesozoic-Cenozoic tectonic evolution of the Colorado River region, California, Arizona, and Nevada: San Diego, Cordilleran Publishers, p. 158–172.

Griggs, D. T., 1967, Hydrolytic weakening of quartz and other silicates: Royal Astronomical Society Geophysical Journal, v. 14, p. 19–31.

Gross, W. W., and Hillmeyer, F. L., 1982, Geometric analysis of upper-plate fault patterns in the Whipple-Buckskin detachment, California, *in* Frost, E. G., and Martin, D. L., eds., Mesozoic-Cenozoic tectonic evolution of the Colorado River region, California, Arizona, and Nevada: San Diego, Cordilleran Publishers, p. 256–266.

Hamilton, W., 1982, Structural evolution of the Big Maria Mountains, northeastern Riverside County, southeastern California, *in* Frost, E. G., and Martin, D. L., eds., Mesozoic-Cenozoic tectonic evolution of the Colorado River region, California, Arizona, and Nevada: San Diego, Cordilleran Publishers, p. 1–28.

Harms, T. A., 1982, The Newport fault: Low-angle normal faulting and Eocene extension, northeast Washington and northwest Idaho [M.S. thesis]: Kingston, Queen's University, 157 p.

Harms, T. A., and Price, R. A., 1983, The Newport fault, Eocene crustal stretching, necking and listric normal faulting in northeast Washington and northwest Idaho: Geological Society of America Abstracts with Programs, v. 15, p. 309.

Haxel, G. B., and Grubensky, M. J., 1984, Tectonic significance of localization of middle-Tertiary detachment faults along Mesozoic and early Tertiary thrust faults, southern Arizona region: Geological Society of America Abstracts with Programs, v. 16, p. 533.

Howard, K. A., Goodge, J. W., and John, B. E., 1982a, Detached crystalline rocks of the Mohave, Buck and Bill Williams mountains, western Arizona, *in* Frost, E. G., and Martin, D. L., eds., Mesozoic-Cenozoic tectonic evolution of the Colorado River region, California, Arizona, and Nevada: San Diego, Cordilleran Publishers, p. 377–393.

Howard, K. A., Stone, P., Pernokas, M. A., and Marvin, R. F., 1982b, Geologic and geochronologic reconnaissance of the Turtle Mountains area, California: West border of the Whipple Mountains detachment terrane, *in* Frost, E. G., and Martin, D. L., eds., Mesozoic-Cenozoic tectonic evolution of the Colorado River region, California, Arizona, and Nevada: San Diego, Cordilleran Publishers, p. 341–355.

Hyndman, D. W., 1980, Bitterroot dome–Sapphire tectonic block, an example of a plutonic-core gneiss-dome complex with its detached suprastructure, *in* Crittenden, M. D., Jr., Coney, P. J., and Davis, G. H., eds., Cordilleran metamorphic core complexes: Geological Society of America Memoir 153, p. 427–444.

Keith, S. B., 1977, The Cenozoic Galiuro and Basin-Range orogenies in southern Arizona: Arizona University, 5th Annual Geoscience Daze, Abstracts with Programs, p. 18.

—— , 1982, Paleoconvergence rates determined from K_2O/SiO_2 ratios in magmatic rocks and their application to Cretaceous and Tertiary tectonic patterns in southwest North America: Geological Society of America Bulletin, v. 93, p. 524–532.

Keith, S. B., and Reynolds, S. J., 1980, Geochemistry of Cordilleran metamorphic core complexes, *in* Coney, P. J., and Reynolds, S. J., eds., Cordilleran metamorphic core complexes and their uranium favorability: U.S. Department of Energy Open-File Report GJBX-258 (80), p. 247–310.

—— , 1981, Low-angle subduction origin for paired peraluminous-metaluminous belts of mid-Cretaceous to early Tertiary Cordilleran granitoids: Geological Society of America Abstracts with Programs, v. 13, p. 63–64.

Keith, S. B., Reynolds, S. J., Damon, P. E., Shafiqullah, M., Livingston, D. E., and Pushkar, P. D., 1980, Evidence for multiple intrusion and deformation within the Santa Catalina-Rincon-Tortolita crystalline complex, southeastern Arizona, *in* Crittenden, M. D., Jr., Coney, P. J., and Davis, G. H., eds., Cordilleran metamorphic core complexes: Geological Society of America Memoir 153, p. 217–268.

Kerrich, R., and Allison, I., 1978, Flow mechanisms in rocks: Geoscience Canada, v. 5, p. 109–118.

Kerrich, R., Fyfe, W. S., Gorman, B. E., and Allison, I., 1977, Local modification of rock chemistry by deformation: Contributions to Mineralogy and Petrology, v. 65, p. 183–190.

Kerrich, R., Allison, I., Barnett, R. L., Moss, S., and Starkey, J., 1980, Microstructural and chemical transformations accompanying deformation of granite in a shear zone at Mieville, Switzerland; with implications for stress corrosion cracking and superplastic flow: Contributions to Mineralogy and Petrology, v. 73, p. 221–242.

Kerrich, R., LaTour, T. E., and Barnett, R. L., 1981, Mineral reactions participating in intragranular fracture propagation: Implications for stress corrosion cracking: Journal of Structural Geology, v. 3, p. 77–87.

Kerrich, R., Rehrig, W. A., and Wilmore, L. M., 1984, Deformation and hydrothermal regimes in the Picacho metamorphic core complex detachment, Arizona: Oxygen isotope evidence: EOS American Geophysical Union (Transactions), p. 1124.

Kerrich, R., Rehrig, W. A., Arima, M., and McLarty, E., 1986, Southwest metamorphic core complexes—Upper plate Tertiary potassic mafic volcanics and their hydrothermal metasomatism: Geological Society of America Abstracts with Programs, v. 18, p. 124.

Lipman, P. W., 1981, Volcano-tectonic setting of Tertiary ore deposits, southern Rocky Mountains, *in* Dickinson, W. R., and Payne, W. D., eds., Relations of tectonics to ore deposits in the southern Cordillera: Arizona Geological Society Digest v. XIV, p. 199–214.

Lister, G. S., 1984, Complexities in the evolution of low-angle crustal shear zones during continental extension: Geological Society of America Abstracts with Programs, v. 16, p. 577.

Lister, G. S., and Davis, G. A., 1983, Development of mylonitic rocks in an intracrustal laminar flow zone, Whipple Mountains, southeastern California: Geological Society of America Abstracts with Programs, v. 15, p. 310.

Logan, R. E., and Hirsch, D. D., 1982, Geometry of detachment faulting and dike emplacement in the southwestern Castle Dome Mountains, Yuma County, Arizona, *in* Frost, E. G., and Martin, D. L., eds., Mesozoic-Cenozoic tectonic evolution of the Colorado River region, California, Arizona, and Nevada: San Diego, Cordilleran Publishers, p. 598–607.

Longwell, C. R., 1945, Low-angle normal faults in the Basin and Range province: American Geophysical Union Transactions, v. 26, p. 107–118.

Lyle, J. H., 1982, Interrelationship of late Mesozoic thrust faulting and mid-Tertiary detachment faulting in the Riverside Mountains, southeast California, *in* Frost, E. G., and Martin, D. L., eds., Mesozoic-Cenozoic tectonic evolution of the Colorado River region, California, Arizona, and Nevada: San Diego, Cordilleran Publishers, p. 470–492.

Mathis, R. L., 1982, Mid-Tertiary detachment faulting in the southeastern Newberry Mountains, Clark County, Nevada, *in* Frost, E. G., and Martin, D. L., eds., Mesozoic-Cenozoic tectonic evolution of the Colorado River region, California, Arizona, and Nevada: San Diego, Cordilleran Publishers, p. 326–340.

Mattauer, M., Collot, B., and Van der Driessche, J., 1983, Alpine model for the internal metamorphic zones of the North American Cordillera: Geology, v. 11, p. 11–15.

McClelland, W. C., 1982, Structural geology of the central Sacramento Mountains, San Bernardino County, California, *in* Frost, E. G., and Martin, D. L., eds., Mesozoic-Cenozoic tectonic evolution of the Colorado River region, California, Arizona and Nevada: San Diego, Cordilleran Publishers, p. 401–407.

Miller, F. K., 1971, The Newport fault and associated mylonites, northeastern Washington: U.S. Geological Survey Professional Paper 750-D, p. D77–D79.

Miller, F. K., and Engels, J. C., 1975, Distribution and trends of discordant ages of the plutonic rocks of northeastern Washington and northern Idaho: Geo-

logical Society of America Bulletin, v. 86, p. 517–528.
Miller, E. L., and Gans, P. B., 1984, Continent transects C1 and C2: Cenozoic extension of the northern Basin-Range province: Geological Society of America Abstracts with Programs, v. 16, p. 596.
Moore, J. G., 1960, Curvature of normal faults in the Basin and Range province of the western United States: U.S. Geological Survey Professional Paper 400-B, p. B409–B411.
Mueller, K. J., Frost, E. G., and Haxel, G., 1982, Mid-Tertiary detachment faulting in the Mohawk Mountains of southwestern Arizona, in Frost, E. G., and Martin, D. L., eds., Mesozoic-Cenozoic tectonic evolution of the Colorado River region, California, Arizona, and Nevada: San Diego, Cordilleran Publishers, p. 448–458.
Nicholas, A., and Poirier, J. P., 1976, Crystalline plasticity and solid state flow in metamorphic rocks: New York, John Wiley and Sons, 230 p.
Okulitch, A. V., 1984, The role of the Shuswap metamorphic complex in Cordilleran tectonism: A review: Canadian Journal of Earth Sciences, v. 21, p. 1171–1193.
Phillips, Jan-Claire, 1982, Character and origin of cataclasite developed along the low-angle Whipple detachment fault, Whipple Mountains, California, in Frost, E. G., and Martin, D. L., eds., Mesozoic-Cenozoic tectonic evolution of the Colorado River region, California, Arizona, and Nevada: San Diego, Cordilleran Publishers, p. 109–116.
—— , 1984, Characterization and deformation mechanisms of cataclastite developed along the Whipple detachment fault, Whipple Mountains, San Bernardino County, California [M.S. thesis]: Los Angeles, University of southern California, 132 p.
Price, R. A., Archibald, D., and Farrar, E., 1981, Eocene stretching and necking of the crust and tectonic unroofing of the Cordilleran metamorphic infrastructure, southeastern British Columbia and adjacent Washington and Idaho: Geological Association of Canada Abstracts, v. 6, p. A47.
Proffett, J. M., Jr., 1977, Cenozoic geology of the Yerington district, Nevada, and implications for nature and origin of Basin and Range faulting: Geological Society of America Bulletin, v. 88, p. 247–266.
Ramsay, J. G., 1980, Shear zone geometry: A review: Journal of Structural Geology, v. 2, p. 83–99.
Ramsay, J. G., and Graham, R. H., 1970, Strain variation in shear belts: Canadian Journal of Earth Sciences, v. 7, p. 786–813.
Rankin, P. W., 1980, Mylonitic fabric development through the east flank of the Bitterroot dome, Montana [M.S. thesis]: Missoula, University of Montana, 80 p.
Rehrig, W. A., 1981, Principal tectonic effects of the Mid-Tertiary orogeny in the Sonoran Desert province: U.S. Geological Survey Open-File Report 81-503, p. 90–92.
—— , 1982, Metamorphic core complexes of the southwestern United States—An updated analysis, in Frost, E. G., and Martin, D. L., eds., Mesozoic-Cenozoic tectonic evolution of the Colorado River region, California, Arizona, and Nevada: San Diego, Cordilleran Publishers, p. 551–560.
—— , 1984, Preliminary geochemical data on the "chlorite breccia" of metamorphic core complex detachment zones and thoughts on associated (?) base and precious metals mineralization: Association of Exploration Geochemists Symposium, Abstracts with Program, p. 23.
Rehrig, W. A., and Heidrick, T. L., 1972, Regional fracturing in Laramide stocks of Arizona and its relationship to porphyry copper mineralization: Economic Geology, v. 67, p. 198–213.
—— , 1976, Regional tectonic stress during the Laramide and late Tertiary intrusive periods, Basin and Range province, Arizona: Arizona Geological Society Digest, v. 10, p. 205–228.
Rehrig, W. A. and Reynolds, S. J., 1980, Geologic and geochronologic reconnaissance of a northwest-trending zone of metamorphic core complexes in southern and western Arizona, in Crittenden, M. D., Jr., Coney, P. J., and Davis, G. H., eds., Cordilleran metamorphic core complexes: Geological Society of America Memoir 153, p. 131–158.
—— , 1981, Eocene metamorphic core complex tectonics near the Lewis and Clark zone, western Montana and northern Idaho: Geological Society of America Abstracts with Programs, v. 13, p. 102.
Rehrig, W. A., Shafiqullah, M., and Damon, P. E., 1980, Geochronology, geology and listric normal faulting of the Vulture Mountains, Maricopa County, Arizona: Arizona Geological Society Digest, v. 12, p. 89–110.
Rehrig, W. A., Halfkenny, R. D., Jr., and Kerrich, R., 1986a, Tectonic-economic significance of isotopic and geochemical data from detachment zones of metamorphic core complexes (MCC), southwestern United States: Geological Society of America Abstracts with Programs, v. 18, p. 174.
Rehrig, W. A., Reynolds, S. J., and Armstrong, R. L., 1986b, A tectonic and geochronological overview of the Priest River crystalline complex, northeastern Washington and northern Idaho, in Schuster, J. E., ed., Selected papers on the geology of Washington: Washington Division of Geology and Earth Resources Bulletin (in press).
Reynolds, S. J., 1982, Geology and geochronology of the South Mountains, central Arizona [Ph.D. thesis]: Tucson, University of Arizona, 220 p.
Reynolds, S. J., and Rehrig, W. A., 1980, Mid-Tertiary plutonism and mylonitization, South Mountains, central Arizona, in Crittenden, M. D., Jr., Coney, P. J., and Davis, G. H., eds., Cordilleran metamorphic core complexes: Geological Society of America Memoir 153, p. 159–175.
Rhodes, B. P., and Hyndman, D. W., 1984, Kinematics of mylonites in the Priest River "metamorphic core complex," northern Idaho and northeastern Washington: Canadian Journal of Earth Science, v. 21, p. 1161–1170.
Rogers, R. D., 1981, Preliminary analyses of the role of fluid pressure in metamorphic core complexes: Geological Society of America Abstracts with Programs, v. 13, p. 103.
Rowles, L., 1982, Deformational history of the Hampton Creek Canyon area, northern Snake Range, Nevada [M.S. thesis]: Palo Alto, Stanford University, 160 p.
Rutter, E. H., 1976, The kinetics of rock deformation by pressure solution: Royal Society of London Philosophical Transactions, v. 283A, p. 203–217.
Scarborough, R., and Meader, N., 1983, Reconnaissance geology of the northern Plomosa Mountains, LaPaz County, Arizona, in Field trip guide to northern Plomosa Mountains, Granite Wash Mountains, and western Harquahala Mountains: Arizona Geological Society fall field trip guide, 1983, 35 p.
Shafiqullah, M., Lynch, D. J., Damon, P. E., and Pierce, H. W., 1976, Geology, geochronology and geochemistry of the Picacho Peak area, Pinal County, Arizona: Arizona Geological Society Digest, v. 10, p. 305–324.
Sibson, R. H., 1977, Fault rocks and fault mechanisms: Geological Society of London Journal, v. 133, p. 191–213.
Smith, R. B., 1983, Cenozoic tectonics of the eastern Basin-Range: Inferences on the origin and mechanism from seismic reflection and earthquake data: Geological Society of America Abstracts with Programs, v. 15, p. 287.
Snoke, A. W., 1980, Transition from infrastructure to suprastructure in the northern Ruby Mountains, Nevada, in Crittenden, M. D., Jr., Coney, P. J., and Davis, G. H., eds., Cordilleran metamorphic core complexes: Geological Society of America Memoir 153, p. 287–334.
Spear, F. S., and Selverstone, J., 1983, Water exsolution from quartz: Implications for the generation of retrograde metamorphic fluids: Geology, v. 11, p. 82–85.
Stewart, J. H., 1979, Regional tilt patterns of late Cenozoic Basin-Range fault blocks in the Great Basin: Geological Society of America Bulletin, v. 91, p. 460–464.
Todd, V. R., 1980, Structure and petrology of a Tertiary gneiss complex, Raft River Mountains, Utah, in Crittenden, M. D., Jr., Coney, P. J., and Davis, G. H., eds., Cordilleran metamorphic core complexes: Geological Society of America Memoir 153, p. 349–384.
Wernicke, B., 1981, Low-angle normal faults in the Basin and Range province: Nappe tectonics in an extended orogen: Nature, v. 291, p. 645–648.
—— , 1985, Uniform-sense normal simple shear of the continental lithosphere: Canadian Journal of Earth Science, v. 22, p. 108–125.
Wernicke, B., and Burchfiel, B. C., 1982, Modes of extensional tectonics: Journal of Structural Geology, v. 4, p. 105–115.
Wilkens, J., Jr., and Heidrick, T. L., 1982, Base and precious-metal mineralization related to low-angle tectonic features in the Whipple Mountains, Cali-

fornia and Buckskin Mountains, Arizona, *in* Frost, E. G., and Martin, D. L., eds., Mesozoic-Cenozoic tectonic evolution of the Colorado River region, California, Arizona, and Nevada: San Diego, Cordilleran Publishers, p. 182–204.

Wright, L. A., and Troxel, B. W., 1973, Shallow-fault interpretation of Basin and Range structure, southwestern Great Basin, *in* deJong, K. A., and Scholten, R., eds., Gravity and tectonics: New York, John Wiley and Sons, p. 397–407.

Zoback, M. L., Anderson, R. E., and Thompson, G. A., 1981, Cenozoic evolution of the state of stress and style of tectonism in the Basin and Range province of the western United States, *in* Vine, F. J., and Smith, A. G., eds., Extensional tectonics associated with convergent plate boundaries: Royal Society of London, 442 p.

MANUSCRIPT ACCEPTED BY THE SOCIETY MARCH 4, 1986

Typeset by WESType Publishing Services, Inc., Boulder, Colorado
Printed in U.S.A. by Malloy Lithographing, Inc., Ann Arbor, Michigan

RAYMOND H. FOGLER LIBRARY
DATE DUE

BOOKS ARE SUBJECT TO